Discover Together: Storytelling for the Whole Family" – Volume 3: Health and Biotechnology

Contents

Foreword

Hello and thank you for visiting "Discover Together: Storytelling for the Whole Family: Volume 3." For me, as a PhD in chemical engineering, the path has been one of ongoing exploration, as I have faced and overcome difficult obstacles while searching for answers in the vast and complicated scientific universe. Along the way, I have gained a deeper appreciation for science and developed a strong desire to teach people about it.

In the course of my professional life, I have had the honor of reading and listening to countless scientific papers and talks, all of which have provided me with valuable knowledge. These events have shown me how seemingly unrelated scientific disciplines are really interdependent on one another, and they have deepened my appreciation for the natural world. I am thrilled to translate and impart to you this treasure trove of information, acquired through much study and reading, in this volume.

In this third part of our series, we explore the fascinating world of health and biotechnology, where state-of-the-art scientific discoveries meet the wonders of human health. By illuminating the miraculous aspects of our biological composition and the groundbreaking achievements in biotechnology, each chapter is meant to do more than just enlighten; it is also meant to inspire. This collection showcases the remarkable human capacity for innovation and indefatigable quest for knowledge, spanning from the intricate workings of the human microbiome to the revolutionary possibilities of artificial organs.

I wish you a sense of wonder and enlightenment as you explore the health and biotechnology fields through these pages, and I hope that your curiosity is piqued along the way. Whether you are an old hand at this or just starting to get interested, I hope you find something new and interesting in this book on the marvels of science.

I appreciate you coming along on this adventure with me. "Discover Together: Storytelling for the Whole Family" is a wonderful resource that I hope will spark a love of learning in all of you, as well as fulfill your natural curiosity and expand your horizons. Cheers to an adventure in discovery that I pray will be just as interesting and rewarding for you as it has been for me.

Chapter 1: Genetic Engineering – Shaping Humanity's Future

Introduction to Genetic Engineering

You have entered a world where the fundamental elements of life are not merely examined but also molded; a world where the boundary between the real and the possible is increasingly blurred by scientific inquiry. This is the realm of genetic engineering, which is both an astonishing and divisive area of contemporary research. Not only are we delving into a scientific field in this chapter, but we are also entering a world that has the potential to reshape our own humanity.

Fundamentally, genetic engineering is the practice of employing biotechnology to alter an organism's genes. From science fiction fantasies to state-of-the-art lab work, this area has seen tremendous growth in recent decades. Looking at its origins can help us comprehend this field better. A number of watershed moments in the history of biology—including the identification of the double helix structure of DNA and the revolutionary discovery of the human genome—have shaped genetic engineering and its trajectory.

Entering the realm of this field's current capabilities is like stepping into a field of exceptional potential. Revolutionary technologies such as CRISPR have broken down previously insurmountable barriers by enabling scientists to modify genes with pinpoint accuracy. Genetic engineering has produced measurable and practical results, not merely theories. From drought-resistant GMO crops to gene therapies that could one day cure hereditary disorders, we have seen incredible progress. Limitations and difficulties accompany these achievements, however, and serve as a constant reminder that our knowledge is still in its infancy.

The ability to alter our genetic code would have far-reaching consequences, particularly if we could design our own children. Potentially altering the course of human health and longevity is the ability to eliminate hereditary disorders prior to a child's birth. On the other hand, it prompts critical inquiries regarding our evolutionary history. If we have the power to select the genes that our offspring inherit, how will our species evolve? When genetic alteration is the standard, what kind of social mores will replace it?

These inquiries touch on both philosophical and ethical dimensions. Having the ability to change the genetic code is a huge responsibility. Our morality and ethics are called into question as we face difficult moral challenges. Is it moral to alter a person's DNA before they are born? Is there a way to prevent the abuse of power or the exacerbated wealth disparity that could result from this kind of technology?

More than simply an overview of the science, this genetic engineering primer asks readers to consider how our species will fare in the future. Along the way, we will investigate not just the facts but also the anecdotes, ethical dilemmas, social effects, and future possibilities of genetic engineering.

Let us, then, go out on this adventure with curiosity and an open mind, prepared to discover the genetic engineering's secrets and potential. With each page turn, we'll learn that genetic engineering is a story about changing the story of humanity, not merely its DNA.

The Science Behind Designing Offspring

As we explore genetic engineering further, we reach the core of its most significant use: the capacity to create progeny. With each new scientific discovery, this idea, which was once only possible in science fiction, is becoming more and more plausible. Delving into this world is a profound exploration of human ambition, ethics, and the essence of existence, going beyond a purely scientific pursuit.

Several advanced tools and methods form the basis of the narrative surrounding the design of progeny through genetic engineering. In the vanguard is CRISPR-Cas9, a game-changing technology that has completely altered our capacity to alter genes. Scientists can now break DNA strands at precise spots with this technology, which is based on a bacterial genome editing system that occurs naturally. Their ability to do so extends to the removal, addition, and modification of DNA sequence segments. It's like a molecular pair of scissors that are directed by a GPS, allowing for exact alterations that have far-reaching ramifications.

However, CRISPR is merely a component. Another tool employed by genetic engineers is gene therapy, which involves modifying an individual's DNA in order to cure or prevent a disease. An strategy that has demonstrated potential in treating illnesses such as hemophilia and specific types of blindness is the insertion of a healthy gene into a person's cells to replace a defective gene.

Amazing things have happened as a result of genetic engineering. There has been and will be tangible progress toward improving food security through the development of pest-resistant crops and the use of genetically modified animals to study diseases that affect humans. Despite these advancements, the discipline is still in its early stages, and there are many obstacles and restrictions to overcome. With every new development, concerns regarding safety, effectiveness, and the consequences in the long run go unsolved.

Beyond repairing genetic flaws, the ability to design kids opens the door to selecting features ranging from appearance to, maybe, IQ or personality. Questions regarding human diversity and identity are fundamentally prompted by this possibility. When we can choose and manipulate individual characteristics, what does this mean for the idea of 'natural' human variation? We must find a way to respect the natural progression of human genetics while also aiming to prevent the misery that genetic illnesses inflict.

Here we delve into more than only the nuts and bolts of possible conception plans for our progeny. We also set out on an ethical and philosophical quest. We take a look at the choices that could determine our species' fate and the obligations that accompany such powerful abilities. The choices we make now could have far-reaching consequences for future generations; we are at a crossroads where science and human agency converge.

This investigation into the field of genetic engineering is more than just an educational exercise; it also prompts us to consider the kind of world we wish to live in. It challenges us to consider the impact of science on our lives and the lives of future generations with both skepticism and empathy. As we consider these possibilities, we realize that we are not merely on the sidelines of science, but rather, we are actively contributing to a story that is still in progress.

Impact on Human Evolution

With the development of genetic engineering, the course of human evolution, which has taken place over millions of years, is about to take a turning point. The possibility of using genetic engineering to create new generations suggests that evolution may one day be influenced by deliberate human action rather than just random chance. In this part of our investigation, we will look at the complex and deep ways in which genetic engineering has the potential to change how humans evolve.

Natural selection and chance mutations in DNA have dictated that evolution has progressed slowly and steadily throughout history. But genetic engineering adds a new variable: the ability to make specific, intentional alterations to our DNA. Because of our newfound ability to alter DNA, we may soon be able to eradicate whole genetic illnesses. Imagine a future where some diseases are a thing of the past, and where evolution of the human race occurs not over millennia, but over generations. The ramifications are immense.

The complicated and mostly unknown long-term genetic consequences of such operations, however, are real. We are effectively seizing control of evolution when we begin to manipulate genes through selection or editing. Worst of all, when we begin to make these decisions, what happens to the variety of the human gene pool? Is it possible that the unintended effects of modifying human DNA would not become apparent until far later in life?

Our actions now have the potential to impact future generations in ways that are difficult to predict or even fathom. Having the ability to influence our evolutionary trajectory is accompanied by a great deal of responsibility. It makes us question our humanity and the variety that makes us unique. Do we want to give up the things that make us special if we live in a society where flawless genes are the standard?

New social standards and expectations may also emerge as a result of the capability to design progeny. Once the stuff of dystopian fiction, the idea of "designer babies" could soon be a reality, allowing users to choose and change a person's physical characteristics, IQ, and talents just like they would a character's traits on a character creation screen. Our conception of human potential would be upended, and societal disparities could be worsened, in such a world. We run the danger of bringing about a society where genetics serve to further divide people if genetic improvements are made exclusively available to the wealthy.

Societal Norms and Ethical Considerations

As we venture further into the realm of genetic engineering, particularly in the context of designing offspring, we find ourselves navigating a sea of complex ethical considerations and shifting societal norms. This part of our journey is not just about the science behind genetic manipulation but delves into the moral and ethical landscape that it reshapes.

The power to design our children at a genetic level brings with it profound ethical dilemmas. One of the most pressing is the question of equity and access. In a world where genetic engineering is possible, who gets to benefit from these advancements? If such technologies are expensive, they could widen the existing gaps in society, creating a divide between the genetically enhanced and those who are not. This disparity could manifest in various forms, from health to cognitive abilities, potentially leading to a new form of inequality based on genetic modification.

Then there is the ethical debate over the extent of human intervention in natural processes. Genetic engineering, especially when it involves designing offspring, treads into areas that have traditionally been left to chance or seen as the realm of a higher power. The concept of 'playing God' is often cited in these discussions, raising questions about the moral implications of such profound control over human life. Where do we draw the line between therapeutic intervention and enhancement? And who decides what constitutes a 'desirable' or 'undesirable' trait?

Moreover, the ability to design offspring challenges long-standing societal norms and values. It could redefine our perceptions of normalcy, health, and beauty. As parents potentially select traits they deem favorable, we might see shifts in societal standards and expectations. This could lead to homogenization, where diversity and uniqueness are undervalued or lost.

The ethical considerations extend beyond individual and societal implications to encompass issues of consent and autonomy. A child whose genome has been engineered has had life-altering decisions made on their behalf before they were even born. This raises questions about their rights and autonomy – did they have a say in their genetic makeup? And should they have had that say?

This section of our exploration is about sparking a dialogue, engaging with the diverse and sometimes conflicting perspectives on the ethical ramifications of genetic engineering. As we traverse this landscape of moral quandaries and societal shifts, we gain a deeper understanding of the responsibilities that come with such profound scientific capabilities. The narrative encourages us to ponder not just the scientific 'how', but the ethical 'should we', inviting readers to reflect on the kind of future we are creating with our technological advancements.

In a world where the lines between nature and technology blur, our collective choices and values will shape the trajectory of genetic engineering and its place in our society. The conversation around the ethical and societal implications of designing offspring is crucial, as it will guide us in navigating this uncharted territory with wisdom, foresight, and a deep respect for the diversity and sanctity of human life.

Case Studies: Theoretical and Real

By exploring the interconnected worlds of genetic engineering's real-world applications and its theoretical underpinnings, this study aims to illuminate the real-world impacts of genetic engineering on society. In this article, we explore possible futures where parents have more control over their children's genetic makeup and think about the social and personal consequences of this trend. When the majority of parents pick for their children specific physical or intellectual traits, it raises concerns about how societal views of success and beauty can alter. Can we rule out the possibility of a highly homogeneous population as a result of this? We are forced to consider the magnitude of the possible impact and long-term consequences of genetic engineering due to these hypothetical scenarios.

To ground our research in the present state of genetic engineering, we use real-world examples at the same time. Using CRISPR technology to treat genetic diseases like sickle cell anemia is one such example. In addition to illustrating the current state of technology and its potential and constraints, these examples provide a grounding in reality that balances our theoretical speculations. The birth of the first gene-edited infants in China, for example, has sparked worldwide debate and controversy and sheds light on the ethical and safety concerns surrounding human embryo editing.

By taking a two-pronged approach, we may better understand the scientific, social, and ethical challenges raised by genetic engineering while simultaneously highlighting the field's immense promise for improving human health and eradicating disease. This essay seeks to provide a nuanced view of genetic engineering that acknowledges both its potential and the weight of responsibility that comes with it. To achieve this, we shall compare actual circumstances with hypothetical ones. This argument has multiple purposes: it educates, it develops critical thinking and ethical reasoning, and it promotes family conversations about the various aspects of genetic engineering. By conducting this thorough analysis, we aim to emphasize the significance of making informed judgments and conducting ethical scientific research in the rapidly evolving field of genetic engineering.

Public Perception and Policy

Here, we explore both hypothetical and real-world situations to shed light on the concrete societal effects of genetic engineering. Our goal in reviewing these case studies is to provide a more nuanced picture of the ramifications and difficulties of creating progeny by bridging the gap between theoretical potential and real-world realities.

Envision a world when parents frequently choose their children's genetic characteristics. The possible individual and societal ramifications are investigated by means of a number of hypothetical situations. Take a culture where parents actively seek out children with specific personality qualities or levels of intelligence as an example. What impact would this have on how society views attractiveness and success? Would it cause the population to become more uniform, with some characteristics being highly valued? By presenting these possibilities, we may delve further into the possible effects of genetic engineering and invite readers to consider the repercussions of these decisions in the future.

The hypothetical situations show what could be, but the real-life examples show what is. In this review, we examine real-world examples of genetic engineering in action, including CRISPR-mediated therapies for hereditary diseases like sickle cell anemia. These instances provide a grounding in reality for the theoretical potential of genetic engineering by illustrating its present strengths and weaknesses.

We also look at cases from the past that caused a stir in the public sphere. The first gene-edited kids in China are a prominent example of a case that has sparked worldwide debate on the morality and viability of this practice. Cases like this are essential for figuring out how people in the scientific and public sectors reacted to the fast development of genetic engineering.

There are two goals to investigating these case studies. To begin with, it highlights the tremendous promise of genetic engineering as a tool for health promotion and illness prevention. Two, it draws attention to the scientific, social, and ethical difficulties that come with such power. We choose these examples to show how important it is to use genetic engineering responsibly while also taking use of its potential for human benefit.

This part serves to educate readers while simultaneously encouraging them to think critically and ethically through its discussion of both hypothetical and real-world circumstances. It encourages more in-depth understanding and awareness of the consequences of these scientific developments by bringing families together to talk about and debate the many facets of genetic engineering. In the dynamic area of genetic engineering, these case examples highlight the significance of well-informed decision-making and ethical scientific conduct.

The Future of Genetic Engineering

We are confronted with a landscape brimming with tremendous ethical concerns and incredible possibilities as we look into the future of genetic engineering, especially in the realm of designing progeny. Looking farther into the future, this investigation probes the larger effects of these innovations on society, ethics, and human evolution, going beyond simple technological forecasts.

A wide range of ambitious and varied forecasts have been put out regarding genetic engineering. In the future, gene-editing techniques will likely be more accessible, efficient, and precise, according to emerging trends. We could see progress in the treatment of complex genetic problems and maybe even try to alter genes to make people live longer, smarter, or stronger. These innovations have the potential to change people's lives and society as a whole.

Possible landmarks in this area may be on par with the first effective treatments for once incurable genetic diseases or the creation of comprehensive ethical standards for genetic engineering. Additionally, we foresee a future where gene-editing tools are accessible to a wider audience, bypassing the need for specialist labs. Even more ambitious "moonshot" initiatives, similar to the Human Genome Project, may reshape genetics in the future.

On the other hand, there will be obstacles along the road to genetic engineering's future. There will certainly be more heated discussions about equity, consent, and humanity as a whole as this technology develops and permeates more and more aspects of our life. A continuous and ever-changing conversation is required since new social problems and ethical quandaries will arise.

This investigation concludes with some thoughts on the vital functions of teaching, public participation, and ethical scientific conduct. We must be well-prepared for a future where genetic engineering is deeply woven into our society and ethics, as well as our scientific and medical practices. A panorama of enormous potential and a sobering reminder of the heavy ethical considerations that come with such power make up genetic engineering's future, which is a mosaic of promise and responsibility.

Here in the future, genetic engineering isn't just a tool for scientists; it's a force that forces us to reevaluate everything from the way we manage health and sickness to the limits of society and ethics. Here on the brink of these revolutionary changes, we are encouraged to face the future with an open mind and a critical eye, embracing the opportunities and threats that this sector presents.

Side Story Subsection: Fun Facts and Pop Culture

There is a lively thread in the complex web of genetic engineering that is weaved from the multicolored threads of pop culture and interesting facts. In this section, we'll take a break from the heavy topics surrounding genetic engineering and look at the more entertaining and approachable sides of the technology that have made it into popular culture.

What an interesting journey genetic engineering has taken as seen via the prism of the media. Both the writers and the viewers of contemporary science fiction have been captivated by the idea of genetic engineering, which has been explored in numerous works of fiction. In some cases, these depictions have shown the public the potential of science in the future, while in others, they have reflected the hopes and anxieties of the general population.

Some of the most compelling and influential depictions of genetic engineering can be found in works of fiction and film. The genetically stratified society in "Gattaca" and the resurrected dinosaurs in "Jurassic Park" have both entertained and made audiences think about the moral and practical consequences of genetic engineering. Although these stories are hypothetical, they have ignited genuine discussions regarding the limits of scientific investigation and the ethical borders that we are prepared to or are not prepared to breach.

Surprisingly, real scientific ideas and developments form the basis of numerous of these fictitious stories. The limits between fiction and reality tend to blur at this juncture, which makes it all the more interesting to figure out what real-life events inspired these science fiction works. One way to better appreciate science fiction for its creative and predictive powers is to familiarize oneself with the science that served as inspiration for these stories.

There are several intriguing and lesser-known aspects about genetic engineering that lend mystery and intrigue to the story. These facts provide a window into the fascinating and varied realm of genetic modification, beginning with the first GMO and continuing with the surprising uses of genetic engineering in sectors like as bioenergy and agriculture.

For an extra dose of intrigue, we investigate the results of numerous polls and studies that show how the general public feels about genetic engineering. How have people's views changed as a result of new research and more visibility in the media? These findings provide a snapshot of how people generally feel about genetic engineering.

Quizzes and thought experiments are added to make this genetic engineering adventure even more intriguing. Readers can put their knowledge to the test and participate in entertaining and instructive hypothetical scenarios based on pop culture in these exercises.

By including this "Side Story" into the larger genetic engineering story, we hope to strike a balance between serious ethical and technical debates and more approachable cultural allusions and interesting facts. This method simplifies genetic engineering for all readers, but notably younger ones, and prompts thought about how science influences our daily lives and the world around us.

Conclusion

As we wrap up our investigation of genetic engineering, especially as it pertains to creating new generations, we are paused for deep contemplation. We have traveled through a field that is both innovative and controversial, exploring its complex science, ethical dilemmas, social effects, and cultural resonances.

Genetic engineering is more than just a scientific enterprise; it reflects our innermost human wants, concerns, and ethical considerations. In this final portion, we reflect on the intricate tapestry we've built, realizing this. The ability to alter DNA and thereby influence our species' destiny carries with it a great deal of weight. It forces us to consider carefully the future we wish to build, not only for our own benefit but also for that of future generations.

One cannot exaggerate the significance of education and conversation in molding this future. We must all take part in well-informed conversations regarding the consequences of these technological developments as we teeter on the edge of major scientific breakthroughs. These discussions shouldn't stay inside the scientific community; they should be a part of a larger public dialogue that includes many viewpoints and beliefs.

Responsible scientific conduct, in which ethical concerns are integral to genetic engineering research and applications rather than an afterthought, is something we stress. In this way, we can be sure that we aren't sacrificing human dignity, diversity, or the value of life in our quest for scientific understanding and competence.

What happens next in the story of genetic engineering will depend on the decisions we make now. We hope that readers will keep thinking, asking, and learning about this exciting topic as we wrap up this book. A genetic engineering narrative is more than simply one of scientific triumph; it is also a story of the insatiable human need to know, the struggle to find balance between scientific progress and morality, and the hope for a better tomorrow.

We don't only relay information in "Discover Together: Storytelling for the Whole Family;" rather, we take readers on an adventure that encourages them to think critically about the world around them and their place in it. As we reach the end of this chapter, we eagerly anticipate the next one, where fantasy and science collide, and the potential is limitless, much like our shared interest.

Chapter 2: Medical Advances – The Future of Global Health and Longevity

Introduction to Medical Advances – The Future of Global Health and Longevity

In the grand narrative of human progress, few chapters are as compelling as the story of medical advancement. As we embark on this journey through the ever-evolving landscape of healthcare and medicine, we find ourselves at the cusp of what could be the most transformative era in the history of human health. This chapter is not just a chronicle of scientific breakthroughs; it's a window into a future where the boundaries of life, health, and longevity are redefined by our relentless pursuit of medical knowledge.

Our tale begins by tracing the footsteps of medical pioneers who laid the foundation of modern medicine. From the discovery of the first antibiotic to the unraveling of the human genome, each milestone in medical history has been a stepping stone leading us to a future brimming with possibilities. These stories of triumph, often born out of necessity and curiosity, have not only shaped the course of medical science but have also profoundly impacted the fabric of society.

As we delve deeper, we find ourselves in the midst of a remarkable era where the conquest of once-invincible ailments seems within our grasp. The audacious goal of curing all forms of cancer by 2050 stands as a testament to this optimism. This vision, ambitious as it is, is rooted in a series of groundbreaking innovations and technologies that have redefined what is possible in the realm of oncology. From precision medicine to revolutionary gene-editing tools, the arsenal against cancer is more formidable than ever.

But the implications of these medical advancements extend far beyond the realm of disease treatment. They weave into the very essence of global health, promising to reshape not only how we treat illness but also how we perceive and experience human life. The potential eradication of cancer could lead to a seismic shift in global health dynamics, affecting everything from healthcare systems to life expectancy across the globe.

This possibility of a longer, healthier life brings us to the fascinating intersection of medicine and society. Increased longevity will undoubtedly have profound implications for societal structures, economies, and personal relationships. How will our societies adapt to a world where living beyond a century is the norm? How will this extended lifespan influence our perspectives on work, education, and intergenerational relationships?

Yet, with great power comes great responsibility. The ability to alter the course of human health and life is laden with ethical complexities and social challenges. As we stand at the frontier of these medical breakthroughs, we must navigate the moral landscape they present. Issues of healthcare equity, the ethics of life extension, and the societal implications of these advancements call for a thoughtful and comprehensive dialogue.

In this introduction to medical advances, we set the stage for a narrative that is as much about human aspiration and ethics as it is about science and technology. This chapter invites readers to embark on a thought-provoking journey through the future of medicine – a journey that not only explores the potential of medical science to conquer diseases but also challenges us to ponder the societal, ethical, and personal dimensions of these advancements.

As we turn the pages of this chapter, we step into a future where the once-impossible becomes possible, where the longevity of life and the quality of health open new horizons for humanity. This exploration is a celebration of human ingenuity and a contemplation of our collective future, shaped by the extraordinary journey of medical advancement.

The Path to Curing All Forms of Cancer by 2050

In the heart of the narrative of medical advancement lies a daring and hopeful promise: the potential eradication of all forms of cancer by the year 2050. This vision, bold and brimming with optimism, is not a mere fantasy but a goal rooted in the remarkable strides made in cancer research and treatment. As we embark on this segment of our journey, we explore the confluence of innovation, perseverance, and scientific brilliance that makes this ambitious goal seem increasingly attainable.

The story of combating cancer is a tapestry woven with threads of breakthroughs and setbacks, of discoveries and relentless dedication. It begins with an understanding of cancer not as a single enemy, but as a multitude of complex diseases, each with its unique characteristics and challenges. The breakthroughs in understanding the genetic and molecular basis of cancer have been pivotal, turning what was once a blanket approach to treatment into a more targeted and personalized strategy.

We delve into the world of modern oncology, where each advancement brings us closer to turning the tide against this formidable foe. From the development of targeted therapies that zero in on specific cancer cells to immunotherapy that harnesses the body's own immune system to fight cancer, the arsenal against this disease is evolving rapidly. The journey is marked by innovative treatments like CAR T-cell therapy, a groundbreaking approach that reprograms a patient's own immune cells to attack cancer cells.

Yet, the path is not without its challenges. Cancer's notorious ability to adapt and evolve makes it a moving target, often requiring a combination of therapies and continuous adaptation of treatment strategies. The battle against cancer is as much a test of scientific ingenuity as it is a war of attrition against a relentless adversary.

The global implications of potentially curing all forms of cancer are monumental. Such a feat would not only signify a triumph over one of humanity's most persistent and devastating diseases but would also herald a new era in global health. The ripple effects would be felt across healthcare systems worldwide, with profound impacts on life expectancy, economic burden, and the overall quality of life. The very fabric of healthcare would be altered, shifting from a focus on cancer treatment to a greater emphasis on prevention and overall well-being.

As we explore this path to curing all forms of cancer, we are not just witnessing the evolution of medical treatments but are also participating in a story of human resilience and hope. This section of the chapter is a tribute to the countless researchers, clinicians, and patients who have contributed to this journey, a journey that stands as a beacon of what can be achieved when science, determination, and the pursuit of a noble goal converge.

In contemplating the future where cancer could be a thing of the past, we open our minds to the limitless potential of medical science and the enduring spirit of human endeavor. This exploration is a testament to our collective dream of a healthier, longer life and a future where the specter of cancer no longer looms over humanity.

Implications for Global Health

The potential eradication of all forms of cancer by 2050 is not just a milestone in medical science; it's a transformative event with far-reaching implications for global health. This part of our journey examines the profound impact that conquering cancer would have on societies around the world, reshaping healthcare systems, altering life expectancy, and redefining our approach to disease management.

Imagine a world where cancer, once a feared and often fatal disease, becomes a manageable condition or even a rarity. Such a monumental achievement in healthcare would significantly alter the global landscape of disease and treatment. Healthcare systems, long burdened by the high costs and complexities of cancer care, would experience a seismic shift. Resources traditionally allocated for cancer treatment could be redirected towards other pressing health issues, potentially leading to a more balanced and efficient healthcare model globally.

The eradication of cancer would also have a profound impact on life expectancy. For centuries, the specter of cancer has been a limiting factor on the human lifespan. With its potential cure, we could witness a remarkable increase in global average life expectancy, opening up new possibilities and challenges in how we manage aging populations. This shift would not only affect healthcare planning but would also have significant economic and social implications, from retirement planning to workforce dynamics.

In countries where access to advanced cancer treatments is currently limited, the eradication of cancer could represent a significant leap towards healthcare equity. It could bridge the gap in health outcomes between developed and developing nations, contributing to a more balanced global health landscape. However, this positive change also brings with it the challenge of ensuring that the benefits of cancer eradication are shared equitably across different regions and socio-economic groups.

Furthermore, the focus of healthcare could shift more towards prevention and overall well-being. With one of the major diseases conquered, public health initiatives might pivot towards preventing other non-communicable diseases, enhancing mental health services, and promoting lifestyle changes that contribute to longer, healthier lives.

The potential impact of a cancer-free world on global health is as broad as it is deep. It invites us to reimagine our approach to healthcare, from the allocation of resources to the training of medical professionals. This section of the chapter not only highlights the bright prospects of this future but also encourages us to contemplate the planning and preparation needed to navigate this new landscape.

Longevity and Society

The prospect of a significant leap in human longevity, particularly in the context of potentially curing all forms of cancer by 2050, opens a fascinating chapter in the story of our society. This exploration delves into the societal implications of increased human lifespan, a future where living well beyond a century could become the norm. Such a dramatic shift in longevity is not just a matter of adding more years to life; it's about redefining the very structure and rhythm of our societal fabric.

With the extension of life expectancy comes a cascade of changes in demographics. The world would witness a growing population of elderly individuals, presenting both opportunities and challenges. On one hand, the wisdom and experience of older generations could become a more prominent and valuable resource. On the other hand, societies would need to adapt to meet the needs of an aging population, from healthcare and social services to housing and transportation.

This increase in lifespan would also have profound effects on the workforce and economy. Traditional career arcs and retirement plans would need reevaluation. People might have multiple careers throughout their extended lives, leading to a more dynamic and diverse workforce. However, this also raises questions about job availability for younger generations and the sustainability of pension systems.

The dynamics of personal relationships and family structures would also evolve. With longer lifespans, the concept of marriage, parenthood, and generational roles might undergo significant changes. Families could span more generations than ever before, altering the traditional roles and interactions within family units. This could lead to richer, more multi-dimensional family experiences, but also require adjustments in family planning and dynamics.

In a society where people live significantly longer, education and lifelong learning would take on new importance. The concept of education as a phase early in life could give way to a model of continuous learning and skill acquisition. This shift would not only be a response to longer careers and changing job markets but also a way to keep individuals engaged and fulfilled throughout their extended lives.

The psychological impacts of increased longevity cannot be overlooked. The prospect of living longer could change individuals' perspectives on life stages, goals, and priorities. It could also bring new mental health considerations, as people navigate the challenges and opportunities of an extended lifespan.

As we examine the impact of increased longevity on society, we are invited to think beyond the traditional paradigms of age and aging. This section of the chapter encourages us to envision a society that not only accommodates but embraces the extended human lifespan. It prompts us to consider the adaptations needed in our social structures, policies, and personal lives to create a world where longer life is an opportunity for enrichment and growth for individuals and society alike.

In this future, longevity is more than just a medical triumph; it's a societal and cultural journey. It's about reimagining our collective future in a world where the boundaries of life are expanded, offering us a new canvas on which to paint the story of our lives.

Ethical and Social Considerations

As we navigate the waters of increased longevity and the potential eradication of cancer, we are met with a complex network of ethical and social considerations. These considerations challenge us to ponder deeply about the ramifications of our advancements in medicine and their impact on the fabric of society. This segment of our exploration delves into the profound ethical dilemmas and social challenges that arise in a world where medical science has dramatically extended human life and conquered one of its greatest adversaries.

One of the most pressing ethical concerns centers around the accessibility and equity of these advanced medical treatments. If we reach a point where cancer can be effectively cured, who will have access to these treatments? The possibility of such medical breakthroughs brings hope, but it also raises the specter of healthcare inequality. There is a real risk that these life-extending treatments might only be available to the affluent or those in developed nations, thereby exacerbating the existing disparities in global health. We must ask ourselves how we can ensure that the benefits of these medical miracles are distributed fairly and justly across all strata of society.

Furthermore, the ethics of life extension itself present a philosophical conundrum. With the ability to significantly prolong life, we must grapple with questions about the natural course of life and death. What are the implications of extending life well beyond what has traditionally been considered a 'natural' lifespan? How will our attitudes towards aging, death, and the stages of life shift in response to these advancements? These questions challenge not only our medical ethics but also our philosophical and spiritual understanding of existence.

The social implications of such profound changes in human health and lifespan are equally significant. A society where people live much longer will need to rethink its approach to many aspects of life, including work, education, relationships, and social welfare. The structures that have governed our lives for generations may need to be reimagined to accommodate this new reality.

Additionally, there is the consideration of the psychological impact on individuals. Living significantly longer lives may affect mental health and emotional well-being. Issues such as the psychological effects of outliving peers, adjusting to multiple life stages, and facing the challenges of extended old age will need to be addressed.

This section of the chapter does not seek to provide definitive answers to these ethical and social questions. Instead, it aims to highlight the complexity of these issues and the importance of addressing them thoughtfully and proactively. As we step into a future where the boundaries of health and longevity are being redefined, we must do so with a keen awareness of the ethical and social responsibilities that accompany such power.

In exploring these ethical and social considerations, we are reminded of the delicate balance between the pursuit of scientific advancement and the preservation of our core values and principles. This journey is a call to engage in meaningful dialogue, to weigh the benefits and challenges, and to navigate this new era of medical science with wisdom, compassion, and an unwavering commitment to equity and justice.

Technological Frontiers in Medicine

In this section of our voyage, we will visit the area of medical technological frontiers, a region where innovation and ingenuity collide to rethink the possibilities of healthcare. This age is defined by a medical technology revolution, in which developments are more than just incremental improvements, but paradigm leaps that challenge our traditional concept of healthcare.

At the center of this change are technologies that were formerly reserved for science fiction but are now becoming a reality in medical laboratories and hospitals. With its ability to interpret large amounts of medical data, provide insights into complex medical issues, and even assist in diagnostic processes, artificial intelligence (AI) is a cornerstone of this new era. The incorporation of AI into healthcare promises to improve diagnostic accuracy, tailor treatment strategies, and change patient care.

Another frontier is customized medicine, which involves personalizing medical therapy to each patient's unique traits. This strategy departs from the one-size-fits-all model by accounting for the genetic, environmental, and lifestyle characteristics that are unique to each individual. Personalized medicine has far-reaching ramifications, perhaps leading to more effective therapies with fewer adverse effects, altering the patient experience.

We also investigate the advancement of innovative medical devices and instruments, ranging from minimally invasive surgical robots to enhanced imaging technology. These advancements improve medical operation precision, shorten recuperation periods, and improve overall patient outcomes. They represent a future in which medical procedures will be less burdensome and more successful.

Emerging technologies in medicine, such as 3D printing, are also discussed. This technique is beginning to show promise in the creation of personalized medical implants, prostheses, and potentially bioprinting human tissues. The applications of 3D printing in medicine are vast, allowing for hitherto inconceivable solutions.

We are aware of the obstacles that these technological frontiers create as we explore them. The rapid pace of technological innovation in medicine raises concerns about the preparedness of our healthcare systems, medical professional training, and the ethical implications of such strong technologies. As we negotiate this new terrain, significant challenges include data privacy, the digital divide, and the possibility for technology to outperform legal frameworks.

This section of the chapter provides an invitation to analyze the technologies' broader impact on society, healthcare, and the human condition, rather than just an investigation of the technology themselves. As we reach the technical boundaries of medicine, we are witnessing a healthcare renaissance, one that promises to transform our approach to medicine and open up new horizons for human health and well-being.

When considering these breakthroughs, we are encouraged to think critically about how we might use new technologies responsibly and ethically, ensuring that they serve to improve, rather than degrade, the quality of human life. This voyage into the medical technological frontiers is a monument to human ingenuity and a reminder of the bright, yet complex, future that awaits in the realm of healthcare.

Case Studies: Medical Miracles and Challenges

This section of our story explores the medical field's successes and failures through a number of case studies. These examples from real life show how medical discoveries are being put into practice, which sheds light on the incredible opportunities and difficult obstacles that come with medical innovation.

A case study is a miniature narrative that depicts a specific incident in the long and winding history of healthcare. To kick things off, we have stories of medical miracles, where innovative treatments, ground-breaking research, or state-of-the-art technology have had a profound impact on patient outcomes. One such story could be about a patient whose terminal cancer was cured by a new treatment, another about a life-saving surgical robot, and still another about a rare disease that was treated with individualized medicine. In addition to praising medical progress, these stories bring attention to the human aspect, showcasing patient stories, committed doctors, and the dogged quest for recovery.

However, there are times when medical advancements are rocky. Case examples that highlight the difficulties encountered in medicine are also examined. Such cases may involve the failure of otherwise effective treatments, the revelation of previously unknown technical constraints, or the emergence of unanticipated problems. Because they teach us important lessons, inspire new ideas, and motivate us to keep becoming better at healthcare, these kinds of cases are fundamental to the history of medicine.

We go through a range of feelings and understandings, from wonder and inspiration to caution and self-reflection, as we investigate these case studies. The incredible accounts give us reason to be hopeful and proud of the progress we have achieved in medicine. Tales of adversity serve as a sobering reminder of the need of modesty and perseverance when confronted with the complexity and unpredictability of the medical field.

These case studies, taken as a whole, show a complex and ever-changing medical environment, beyond the scope of any one story. They mirror the challenges of combining innovation and safety in clinical practice and the complexities of translating scientific findings. These stories highlight the fact that progress in medicine is a human story about tenacity, creativity, and the never-ending quest for better health outcomes, rather than only a scientific one.

We are encouraged to consider the larger ramifications of these medical stories as we progress through each case study. Every new development and obstacle makes us think about how healthcare will evolve in the future and how it will affect people's lives generally. These case studies are more than just tales; they form the foundation of contemporary medicine's intricate web of knowledge.

Our hope is that by sharing these accounts of medical wonders and tragedies, we might provide a more nuanced picture of medicine, one that recognizes its achievements while also recognizing its shortcomings. This investigation serves as a memorial to the never-ending pursuit of medical brilliance and a timely reminder of the teamwork needed to adapt to the dynamic healthcare system.

The Future of Healthcare and Policy

As we go deeper into the investigation of medical innovations, our attention turns to the healthcare system of the future and the critical role that policy will play in molding it. This section of the chapter is an introspective look into the future of medicine, taking into account the possible innovations, the changing obstacles, and the policy frameworks that will be crucial for overcoming these obstacles.

In this story, healthcare is seen as a field that will undergo constant change and innovation in the future. Here, new therapies, technologies, and ways of thinking about health and wellbeing are always emerging as medical science pushes the frontiers forward. Personalized medicine, in which a patient's treatment is based on their unique genetic composition in addition to their lifestyle and environmental factors, may become widely used in the not-too-distant future. Potentially improving diagnostic precision and treatment outcomes, AI and ML might soon find their way into routine clinical practice.

But new considerations and difficulties accompany these innovations. We must ask if our systems are prepared to keep up with the lightning-fast rate of technology advancement in healthcare. In light of emerging medical practices, how will healthcare systems change to support these innovations? In what ways will this affect the education, experience, and work of healthcare workers?

In this future, policy is crucial. The safe, ethical, and equitable realization of healthcare innovations depends on effective policymaking. The ethical concerns surrounding genetic editing and artificial intelligence in healthcare are only two of many topics that will require policy attention. Making regulations that encourage innovation while protecting public health and patient rights is going to be a tough nut to crack.

The significance of international cooperation and the harmonization of policies is also becoming more apparent when we consider the future. In order to tackle global health issues, share knowledge, and establish common standards, international cooperation is important. This is because the possibilities and challenges of medical developments know no bounds.

Medical innovation, social needs, and policy frameworks will interact dynamically in the future of healthcare, as this section of the chapter invites readers to imagine. It challenges us to reflect on how we can work together to create a healthcare system that is accessible, ethically sound, and technologically capable of the future.

Thinking about this future brings to mind the fact that healthcare is a never-ending adventure, full of new discoveries, obstacles, and opportunities. Scientists, medical experts, lawmakers, and the general public must all work together on this path. Our vision for the future of healthcare is to be an active participant in building a system that is inventive, inclusive, and sensitive to the needs of all people, not merely an observer of the wonders of medical research.

Side Story Subsection: The Fascination with Cryonics and Human Preservation

As we delve into the world of medical discoveries, a fascinating side tale unfolds, one that captivates the imagination and raises profound concerns about the meaning of life and its preservation. The concept of human preservation through cryonics is at the intersection of science and the fantastical world of the future. The concept of cryonics, which involves storing people at very low temperatures in the hopes of their eventual resurrection, has long piqued curiosity and sparked doubt.

The fascinating story starts with the origins of cryonics. From a far-fetched concept in the middle of the twentieth century to the founding of the first cryonics institutions, we follow its history. These institutions symbolize an unwavering belief in the power of future medicine and an intense longing to overcome death by providing a means to prolong life until a cure is found.

Cryonics has stirred up quite a stir, influencing both the scientific community and the general public, as we shall see in the following sections. Cryogenic preservation has been shown in several media as a means for characters to awaken to a new realm of potential. The fictitious nature of these depictions belies our innate interest in exploring the limits of medical knowledge and life extension. Their existence begs the questions of who we are, how our minds work, and the nature of death itself.

Insights into the actual events that served as models for these works of fiction are also provided in this section. We take a look at cryopreservation as a science and talk about how it's being used in modern medicine for things like fertility treatments and organ donation. Concerning the ethical issues that come up from such initiatives, we investigate the limits and difficulties of using this technology to whole-body preservation.

In addition, we provide fascinating information about cryonics, which brings a sense of awe to this intricate subject. Tales of notable people with an interest in cryonics and the unexpected connections between cryonics and other disciplines and industries are among them.

We incorporate quizzes on the subject and hypothetical scenarios influenced by popular culture to make this cryonics trip more interesting and engaging. In addition to learning about cryonics, readers are encouraged to actively engage with the philosophical and ethical concerns it raises through these interactive components.

A interesting new layer to our examination of medical developments is this side tale of cryonics and human preservation. The text encourages readers to think on the strengths and weaknesses of medical technology, both present and future. The eternal human desire for immortality and the dynamic conversation between science, ethics, and imagination are brought to light as we consider the possibilities of cryonics.

Conclusion: A Vision for the Future

As we near the end of this journey into the realm of future medical innovations, we are prompted to contemplate deep thoughts. This exploration of cancer treatment, life extension, and future technological wonders has been more than just a theoretical exercise; it has been a voyage through the opportunities and threats that comprise the essence of our shared human experience.

Here we pause to collect all the story threads and weave them into a healthcare and society vision for the future. This hopeful vision is based on the remarkable medical advancements we have accomplished, but it is also tempered by the knowledge of the duties that accompany this level of authority. In this potential future, cancer may no longer be a specter over people's lives, new chapters in our individual and collective narratives are opened by increased lifespan, and medical technology keeps pushing the limits of what was previously believed to be achievable.

Having said that, there are a number of difficulties to this concept. We are cognizant of the fact that these medical breakthroughs bring with them societal changes, policy hurdles, and ethical dilemmas. Striking a balance between embracing innovation and sustaining the essential values of equality, ethics, and compassion will be crucial for the future of healthcare. A firm dedication to the ideals of fairness and accessibility, in addition to scientific and technological acumen, will be required.

In this imagined future, the importance of teaching, talking, and working together becomes critical. We need to keep talking about the ramifications of medical science as it progresses. Everyone from lawmakers to ethicists to the general public should be a part of these conversations, not just the medical community. In order to tackle the common healthcare issues and make sure that people from all walks of life may benefit from medical advances, it will be crucial for people from all over the world to work together.

We are reminded of the strength and creativity of the human soul as we think about this future. Progress in medicine is a story of how far we've come in our quest for understanding and how determined we are to make people's lives better. With the knowledge we've gained from the past and the tools we've invented, this story encourages us to look ahead with hope.

As we wrap off this chapter, we want to encourage readers to keep up with the dynamic tale of medicine. This is just the beginning of the adventure that science, ethics, and society will embark on together. Every one of us has a part to play in determining the shape and color of healthcare's future, which is like a blank canvas. We move forward, fueled by understanding, empathy, and the common goal of leaving a better, healthier planet to future generations, with an eye toward a future when medical innovations are used to benefit everyone.

Chapter 3: Nano-robotics in Medicine – Revolutionizing Healthcare

Introduction to Nano-robotics in Medicine

Emerging from the enormous and dynamic field of medical science is an intriguing and intimidating new frontier, offering a future that was previously unimaginable. Behold, the realm of nano-robotics in medicine—a realm where the microscopic meets the medicinal—wherein microscopic machines unveil enormous opportunities in healthcare. As we explore into the fascinating realm of nano-robotics, we uncover a narrative that encompasses both technological progress and a revolutionary change in our perspective on health and illness.

An astonishing story of convergence unfolds in the origins of nano-robotics in medicine, as the lines between engineering, physics, biology, and medicine blur to produce something truly remarkable. At its core, nano-robotics is concerned with the design and control of mechanical systems on a size so microscopic that it is measured in nanometers, or one billionth of a meter. To better understand this, think of a nanorobot as being similar to an ant in comparison to a human. The extraordinary accuracy and therapeutic potential offered by these minuscule robots constitute a paradigm shift in the field of medical science.

A brief overview of the evolution of this discipline serves as the foundation of our narrative. A tale of human inventiveness and scientific advancement, the history of nano-robotics spans from the initial hypotheses and notions put out by scientists and visionaries to the initial experimental steps towards the creation of molecular machines. Along the way, we've made strides that defied our wildest imaginations and united experts from all walks of life in a remarkable display of teamwork and invention.

Investigating the present status of nano-robotics in healthcare transports us to a reality where these microscopic machines are more than just ideas. They are being used more and more in a variety of medicinal applications, and they are real and practical. To accomplish things that would be inconceivable with larger robots, researchers are now developing nano-robots that can navigate the human body, detect and target illness cells, and deliver medications with precise accuracy.

There is a wide range of possible medical uses for nano-robots. To maximize therapy success and minimize side effects, picture a swarm of cell-sized nano-robots making their way through the bloodstream to target a malignant tumor. Imagine nano-robots operating on damaged tissues or arteries at the cellular level to fix them or unclog them. Treatments that are less intrusive, more effective, and personalized to each patient's needs are within reach, thanks to the infinite potential of the human body.

However, there will be obstacles along the way. Important concerns and considerations arise with the development and deployment of nano-robots in medicine. What measures can we take to guarantee that these tiny devices are both safe and effective? When we have the means to alter the human body on such a basic level, what moral questions come to light? How can we get these breakthroughs from the lab to the patient while navigating the regulatory landscapes?

This introductory guide on nano-robotics in medicine will take you on a journey into a frontier where science fiction and medical reality converge. This investigation not only brings attention to the technological wonders and possibilities of nano-robotics, but it also prompts serious contemplation on how these developments may affect healthcare in the years to come. Entering this minuscule realm ushers in an era when nano-robotics' limitless potential redraws the lines of healthcare and where the tiniest machines hold the answer to our most pressing medical problems.

The Science of Nano-robotics

As we explore the field of nano-robotics in healthcare further, we are standing at the crossroads of complex science and revolutionary technology. Fundamentally, nano-robotics is a concerto for the exceedingly microscopic, a discipline whose precision is measured in nanometers and whose prospective influence is nothing short of monumental. The science underlying these microscopic wonders, including their operation and the elements that give them life, is the focus of this part of our story.

Manipulating matter at the atomic and molecular size to build machines that can execute particular jobs is fundamental to nano-robotics. From identifying and removing sick cells to mending injured tissues, these jobs are incredibly precise. A wide variety of materials, selected for their specific characteristics and potential biocompatibility with humans, constitute the building blocks of nano-robots. Depending on their intended use, nano-robots can either be made of biocompatible metals or organic materials such as DNA or proteins.

In order to comprehend the inner workings of nano-robots, one must go into the world of microscopic mechanical systems and molecular motors, which tests the limits of our traditional knowledge of mechanics. A number of mechanisms enable these nano-scale devices to move; some use the body's natural chemical energy, while others rely on external magnetic fields or light. Using a combination of biological, chemical, and physical principles, engineers have created nano-robots that can navigate and operate inside the human body.

The incredible degree of control and accuracy offered by nano-robotics is one of its most intriguing features. In contrast to more conventional medical procedures, nano-robots are able to navigate the intricate human body with remarkable precision and little disruption to normal bodily functions. Thanks to this level of accuracy, a plethora of new medical opportunities have emerged, such as nano-sensors that can identify diseases in their early stages and tailored drug delivery systems that may directly treat afflicted areas.

The healthcare applications of nano-robotics as they exist now are in their infancy. Use cases that would have been unimaginable even a generation ago are becoming commonplace. A variety of nano-robot applications are now under development, including the removal of arterial plaque, the targeting of cancer cells with chemotherapy, and the detection and neutralization of viruses. The inventiveness and promise of nano-robotics in transforming healthcare are demonstrated by each application.

The path of nano-robotics, however, is about more than simply the mechanics and science. Perseverance and vision are also significant themes. The development of these little devices is a significant step forward in medical science, fueled by the ambition to expand the limits of healthcare possibilities.

An interesting and potentially fruitful field of medical study is our starting point as we delve into the science of nano-robotics. The trip takes us on an awe-inspiring tour of the nano-world, which holds immense promise for revolutionizing healthcare. This universe ushers in a new age in healthcare where the smallest innovations have enormous effects, and the smallest machines contain the key to some of the greatest discoveries.

Applications in Disease Treatment

As we delve deeper into the realm of nano-robotics, we find a landscape brimming with possibilities for the treatment of diseases. The genuine revolutionary potential of nano-robotics is revealed in this domain, where the tiniest robots encounter the gravest medical problems. In this section of our journey, we explore how nano-robots are being developed and used to transform the way diseases are treated, bringing new possibilities and answers that were previously out of our reach.

Targeted drug delivery is a key idea behind these applications. Common, older approaches to dosing medications have a more blanket-like impact, influencing the entire body and perhaps causing unwanted side effects. On the other hand, nano-robots provide pinpoint accuracy in medicine administration by transporting drugs straight to the cells or tissues that need them. Picture microscopic machines, no larger than a few molecules, floating through the circulatory system, ready to unleash their medicinal cargo precisely where the illness is manifesting, be it an infection or a malignant tumor. This precision medicine is a quantum leap forward in patient care since it increases therapeutic efficacy while decreasing negative effects.

With nano-robots, new possibilities in tissue healing and surgery open up. When it comes to surgical procedures, nano-robots have the potential to provide unmatched precision. They can execute complex cellular-level processes that are unimaginable for human hands or even traditional surgical instruments. In the future, these tiny surgeons may be able to fix broken tissues, unclog arteries, or even put together molecular structures without ever leaving the body.

Among the most insurmountable health problems of our day, nano-robots provide a glimmer of hope in the fight against cancer. In order to target cancer cells and eliminate them while sparing healthy ones, scientists are creating nano-robots. While some of these devices are being developed to essentially starve tumors by cutting off their blood supply, others are being evaluated to inject deadly dosages of chemicals straight into the tumors' cores.

There is promise and creativity in every narrative of using nano-robots to treat sickness. In addition to focusing on the elimination of diseases, these stories also aim to restore health and improve the quality of life. They herald a time when diseases are no longer considered incurable and when therapies are tailored to each individual patient.

We are also cognizant of the fact that there will be obstacles to overcome as we investigate potential applications. Many ethical, legal, and scientific challenges await those who venture from the realm of laboratory research to that of clinical application. The hope for a future when nano-scale solutions, rather than brute force, can tackle the most complicated medical issues is what drives the field of nano-robotics forward, as is the promise of this technology to revolutionize illness treatment.

Advancements in Diagnostic Procedures

Now that we've covered the revolutionary realm of nano-robotics in healthcare, let's move on to the significant effect it will have on diagnostic processes. Here, nano-robots are paving the way for early disease detection and monitoring in addition to being useful therapeutic instruments. Improved diagnostics, earlier interventions, and individualized treatment plans are just a few ways in which nano-robotics is changing the face of healthcare.

A major departure from traditional approaches is the incorporation of nano-robots into diagnostic processes. Getting a diagnosis of an illness, particularly at an early stage, has traditionally been difficult and often required intrusive procedures and complicated testing. But nano-robots provide a fresh perspective. Picture this: all throughout the human body, microscopic sensors (not much bigger than a few cells) would be actively searching for biochemical markers of impending illness. These tiny watchdogs can pick up on abnormalities as little as minute chemical shifts or as early as the first phases of cellular transformation, which are invisible to more conventional diagnostic methods.

Early illness diagnosis and continuous monitoring is an area where nano-robots show the greatest promise as a diagnostic tool. Illnesses including cancer, heart disease, and neurological abnormalities could be detected well in advance of their more serious manifestations. Patients who are at risk might be monitored continuously with these nano-sensors, who could provide information about their health status in real-time. By moving the emphasis from treatment to prevention, this degree of monitoring and accuracy in diagnostics has the potential to radically alter the management of diseases.

Similarly far-reaching is the effect of nano-robots on individualized healthcare. Medical treatment can be better catered to each patient's specific needs with the use of nano-robots, which can collect precise data on an individual basis. In the future, doctors may be able to use this data in conjunction with AI and sophisticated analytics to create individualized treatment programs that maximize the efficacy of medical interventions.

As we explore the diagnostic breakthroughs made possible by nano-robotics, we also come across several hurdles and factors to consider. Thorough testing is necessary to ensure the precision and dependability of these nano-scale diagnostic instruments, and concerns regarding patient confidentiality and data protection must be taken into account. There will be new protocols and medical staff training needs as a result of incorporating such sophisticated technologies into current healthcare systems.

In this part of our investigation, we not only admire the innovative spirit, careful study, and ethical concerns that have led to the remarkable breakthroughs in nano-robotics, but we also marvel at their ability to revolutionize diagnostics. The use of nano-robots in diagnostics demonstrates a commitment to better patient care and an insatiable need for knowledge. It depicts a future when nano-scale devices silently and precisely solve the human body's secrets, rather than invasive treatments, ushering in a new era of medicine where early detection and individualized care are not mere aspirations, but actualities.

Overcoming Healthcare Challenges

As we explore the groundbreaking field of nano-robotics in healthcare, it becomes clear that these small robots can do more than improve existing medical practices; they can also solve some of the biggest problems in healthcare today. Here we go into the ways nano-robotics can help with problems like worldwide health inequities, delayed diagnoses, and medicines that aren't easily accessible.

The capacity to provide answers that conventional medicine fails to address is a distinguishing feature of nano-robotics. For instance, nano-robots could open up new possibilities for diagnosis and treatment in places with limited access to modern medical facilities or few healthcare resources. These little gadgets can provide high-quality medical treatments to underserved or far-flung locations without requiring massive infrastructural upgrades. One promising approach to expanding access to healthcare is nano-robotic technology, which is both scalable and highly portable.

The prompt and precise detection of illnesses is another major obstacle in healthcare. One potential answer is the use of nano-robots, which can roam the human body and detect molecular abnormalities. They are able to detect diseases in their early stages, much before traditional approaches would, which means that therapies can be administered earlier and, perhaps, more effectively. This capacity has the potential to greatly improve patient outcomes by reducing diagnostic delays, particularly in situations when early action is critical.

When it comes to medical care, nano-robots open up new possibilities for treating illnesses that have hitherto proven intractable. When it comes to treating diseases like cancer, where conventional treatments like chemotherapy can cause serious side effects since they don't target specific cells or tissues, their accuracy and specificity are priceless. By avoiding healthy cells and delivering medications straight to malignant ones, nano-robots can lessen the side effects of cancer therapy.

Also, healthcare could become more efficient and tailored with the introduction of nano-robotics. Medical interventions can be made more effective with the use of nano-robotic treatments that are tailored to each patient's unique demands. When it comes to managing long-term health conditions, this level of customization is even more pronounced; nano-robots may track a patient's vitals in real-time and adjust their therapy accordingly.

Many obstacles stand in the way of fully incorporating nano-robotics into conventional healthcare. The creation and implementation of these devices present a number of challenges, including technological ones, as well as those related to regulation and ethics. It is critical that scientists, healthcare providers, and lawmakers work together to ensure that nano-robots used in medicine are safe and effective, and that these discoveries can make it through the regulatory process to the public.

Ethical and Safety Considerations

As we go into the exciting world of nano-robotics in healthcare, we reach a critical point where ground-breaking innovation meets the accompanying ethical and safety concerns. As we continue to explore the potential applications of these tiny technologies in healthcare, we will inevitably encounter serious ethical concerns and obstacles to patient safety. Although nano-robotics holds immense promise, it also forces us to face a tangled web of ethical and pragmatic dilemmas.

There are many different angles from which to view the ethical implications of nano-robotics. Autonomy and consent are at the heart of one of the main issues. Concerns with patient permission emerge due to the fact that nano-robots function on a scale that is incomprehensible to humans. Since the interventions in question are not readily apparent to the human eye and have their basis in highly developed scientific principles, how can we guarantee that participants have given their informed consent?

Misuse of nano-robotic technology is another possible ethical concern. The size, accuracy, and molecular manipulation capabilities of nano-robots—the very qualities that make them so valuable in medicine—also give rise to worries over their application outside of medicine. As part of the ethical conversation around this technology, we must address the risk that nano-robots could be utilized for malicious objectives like surveillance or as biological weapons.

Priority number two is ensuring everyone's safety. Thorough testing and validation are prerequisites for introducing any new medical technology. Making sure nano-robots are harmless when used on humans, easy to manage, and not able to do harm by accident are all part of this. This necessitates a great deal of time, money, and regulatory clearance for research and experiments.

Another intricate layer is the regulatory environment around nano-robotics in healthcare. The subtleties of nano-robotic technology may be too much for the current medical laws to manage. New frameworks to regulate the usage, deployment, and oversight of nano-robots in healthcare settings require the combined efforts of researchers, legislators, and ethicists.

This part of our story does not intend to give final solutions to these safety and ethical questions; rather, it wants to stress the significance of continuing the conversation and being vigilant. Although nano-robotics hold great promise, we must not lose sight of the weight of duty that accompanies their use.

Investigating the potential risks and ethical implications of using nanorobotics in healthcare is an important first step in this exciting new area. To guarantee that the breakthroughs in nano-robotics are achieved in an ethical and safe manner, it calls on all parties involved, including researchers, medical professionals, ethicists, government officials, and patients, to work together. As we move across this landscape, we are constantly reminded of the fine line that must be drawn between venturing into unexplored scientific territory and maintaining the most rigorous standards of safe and ethical medical practice.

Impact on Healthcare Systems

More research into medical nano-robotics points to the impending arrival of these microscopic devices as game-changers in healthcare systems around the globe. This part of our story delves at how nano-robotics will change the face of healthcare delivery and management, altering the parameters of medical treatment and patient care in profound ways.

More accurate, efficient, and individualized healthcare is on the horizon thanks to the integration of nano-robotics into healthcare systems. The use of nano-robots in medicine holds great promise for a number of reasons, including the potential to enhance treatment outcomes, shorten recovery times, and lessen invasiveness of treatments by targeting certain cells and delivering medications directly to affected areas. The advent of precision medicine has ushered in a new age where therapies are customized to meet the unique requirements of each patient, marking a departure from the traditional "one size fits all" mentality.

The introduction of nano-robotics presents healthcare providers with new possibilities as well as new obstacles. While it presents innovative approaches to patient care, it also calls into question long-established norms in medical education and practice. In order to improve patient care in a future where nano-robotics is integral, medical practitioners will have to learn new things and adjust to a new way of thinking about healthcare.

The administration and structure of healthcare institutions are also influenced by nano-robotics. In order to accommodate nano-robotic technology, healthcare facilities will have to undergo a period of transformation. To manage the massive volumes of data produced by nano-robotic diagnostics and treatments, it is necessary to incorporate state-of-the-art data management systems in addition to the more tangible components, such as specialized equipment and facilities.

It is also possible that more people will be able to afford cutting-edge medical treatments if nano-robotics are widely used in healthcare systems. This technology has the potential to lower healthcare access obstacles as it gains popularity and lowers prices, making it available to more people. This has the potential to bring about a more fair healthcare system in which people have equal access to cutting-edge treatments rather than just a select few.

Nevertheless, there are several obstacles to overcome when incorporating nano-robotics into healthcare systems. Fair access to new technologies, adequate financing, and the distribution of available resources are all pressing concerns that must be resolved. For the healthcare business, lawmakers, and stakeholders to overcome these obstacles and reap the benefits of nano-robotics, they must collaborate.

Case Studies: Breakthroughs and Setbacks

We will examine many case studies that showcase the inventive successes and challenging setbacks of nano-robotics in healthcare as we continue our investigation of the area. These practical cases illustrate the potential and limitations of nano-robotics and shed light on the multifaceted impact of this technology on healthcare.

Both the theoretical and practical aspects of medical nano-robotics are enhanced by the narrative style of each case study. We begin with news of groundbreaking developments, wherein diagnostic procedures, patient outcomes, or illness therapy have all been substantially enhanced by nano-robotic technology. In situations where more traditional approaches had failed, a full recovery may have been achieved by employing nano-robots to precisely target and remove cancerous tumors. However, there is new hope for the treatment of cardiovascular disease because of stories of nano-robots that can remove arterial plaque.

However, nano-robotics in healthcare face the same challenges as any other industry when it comes to innovation. Additionally, case stories that demonstrate the challenges and obstacles encountered are examined. Such instances may arise in the course of nano-robotic therapy if the desired results were not attained, if technological limitations were found, or if unforeseen complications emerged. Looking at these kinds of examples can tell us a lot about the current state of nano-robotic technology and what needs correcting.

Examining these case studies allows us to see the pros and cons of utilizing nano-robotic technology in healthcare. These advancements suggest a promising future where nano-robots play a crucial role in healthcare, delivering therapies that are more targeted, less invasive, and more effective. The difficulties show us how crucial it is to carefully test and develop new technology, how complex medical science is, and how to be modest and meticulous.

The fact that these case studies involve the collaboration of engineers, physicians, researchers, and ethicists further demonstrates the interdisciplinary nature of nano-robotics. They highlight the significance of interdisciplinary healthcare innovation, where experts from many fields collaborate to advance medical knowledge.

Exploring these case studies of triumphs and setbacks reveals a journey of unwavering determination, innovative problem-solving, and the ability to embrace and learn from errors in the field of nano-robotics in medicine. Beyond merely detailing technological accomplishments, these stories demonstrate the human desire to better the well-being and quality of life for everybody through the exploration of uncharted territories.

Side Story Subsection: Nano-robotics in Popular Imagination

As an interesting side note to our main article about nano-robotics in healthcare, we investigate the intriguing representations of this cutting-edge technology in science fiction and the media. While nano-robotics has its foundations in science, it has also captivated the imaginative and creative faculties of filmmakers, storytellers, and futurists, as this excursion into popular imagination reveals.

Nano-robotics has been a mainstay of science fiction for quite some time, frequently depicted as a futuristic technology with boundless potential. The public's view and comprehension of nano-robotics have been influenced by many depictions, such as the hordes of microscopic robots in Michael Crichton's "Prey" and the nanobots seen in several science fiction films and television series. In their future, they imagine, small robots can do more than just fix illnesses; they can also improve human capabilities, fix environmental damage, and even break beyond the limits of conventional physics.

While these stories are obviously works of fiction, they frequently mirror the expectations and worries people have about technological progress. Concerns regarding human augmentation, the ethical usage of nano-robots, and the dangers of technology that is perhaps too advanced for human safety are all brought up by these issues. By delving into these topics, popular culture encourages a wider audience to contemplate the philosophical and ethical quandaries that new technologies like nano-robotics pose.

Along with these made-up examples, this side tale also reveals interesting and entertaining facts about nano-robotics in media. As an example, nano-robotics has been present in mainstream conversations about the future of technology, has an impact on video games and virtual reality experiences, and is even used in blockbuster movies. All of these things highlight how fundamental nano-robotics is to our shared imagination and how it influences our hopes for the future.

We add elements like movie and book quizzes on nano-robotics, discussions on how well these depictions reflect the actual state of the technology, and hypothetical situations influenced by popular culture to make this examination of nano-robotics in popular imagination even more interactive and interesting. Not only may readers gain a better understanding of nano-robotics through these interactive components, but they can also actively participate in exploring the philosophical and ethical issues that this field raises.

Our investigation of nano-robotics in healthcare gains a fascinating new dimension from this subplot involving nano-robotics in popular culture. By bringing together factual scientific information with imaginative conjecture, it encourages readers to consider the advantages and disadvantages of this technology. The ability of the human intellect to foresee potential outcomes and consider the far-reaching effects of our technological endeavors is emphasized as we make our way through these stories, some based on reality and others on fiction.

Future Directions and Potential

As we near the end of our investigation into medical nano-robotics, we look ahead, wondering where this revolutionary science may go from here and what possibilities lie ahead. This part provides a vision for the future that is both creative and realistic, reflecting on current trends and the potential revolutionary effects of nano-robotics on healthcare and society.

As nanotechnology, materials science, and biomedical engineering continue to progress, the potential applications of nano-robotics in healthcare are endless. More advanced nano-robots, able to perform ever-more-complicated jobs, are probably on the horizon. Potential examples include nano-surgeons who operate on individual cells and nano-pharmacists who work within the body to produce medications as needed using data collected from the patient's vital signs in real time.

The possibility that nano-robots may cause a sea change in the way chronic diseases are treated is one of the most promising future developments. Nano-robots have the ability to keep tabs on a patient's vitals all the time, alerting doctors to possible problems and allowing them to administer precise treatments before they get worse. Reducing the burden of chronic diseases could be achieved by this proactive approach to healthcare, which focuses on prevention and early treatment.

Future nano-robots may provide even more thorough and accurate monitoring in the field of diagnostics. Envision nano-sensors that can identify genetic or molecular markers of illness years before any outward symptoms appear. These innovations have the potential to change the way diseases are identified at an early stage, allowing more individuals to take an active role in their health management.

We may also anticipate a greater incorporation of nano-robotic technology into preexisting medical processes and practices as this field develops further. As a result of this integration, new medical protocols and methodologies will emerge, and healthcare providers will need to learn how to deal with the intricacies of nano-scale technology.

Nevertheless, there will be obstacles along the way to these future possibilities. Complex ethical considerations, legal impediments, and technological challenges will need to be overcome on the road ahead for nano-robotics in medicine. Ethical debate and deliberate policymaking must keep up with the rate of technological development.

Additionally, it will be critical to guarantee equal access to these cutting-edge treatments as nano-robotics grows in popularity in healthcare. Making sure that everyone can enjoy the benefits of nano-robotics will require finding a way to close the gap between what technology can do and how easy it is to implement.

We conclude by imagining a future when nano-robotics is essential to healthcare advancement. We are all committed to better health outcomes and quality of life, and this future is a reflection of that. It is also a tribute to human inventiveness and technical capability. We feel both hopeful and responsible about this future as we think about how our decisions now will affect the influence of nano-robotics in the years to come. To ensure that even the tiniest robots have a profound impact on healthcare in the future, this investigation of possible paths forward is an impassioned plea for ongoing innovation, careful ethical thinking, and international cooperation.

Conclusion

As we conclude our investigation into the world of nano-robotics in medicine, we are on the approach of a new era in healthcare. This conclusion is more than just a wrap-up; it is an opportunity to look ahead and contemplate the far-reaching repercussions and game-changing prospects of nano-robotics in the realm of medicine and healthcare. This is an open invitation to create a future in which technology and medicine merge to revolutionize healthcare.

In this imagined future, nano-robotics plays a critical part in medical developments, delivering unparalleled levels of efficiency, personalization, and precision in diagnosis and treatment. We envision a future in which disease detection and treatment are routine, medical interventions are precise and polished, and chronic condition management is integrated into people's daily lives.

As we consider the future, we are well aware of the obligations and difficulties that come with such significant technical breakthroughs. On the route ahead for nano-robotics in medicine, ethical challenges, safety concerns, and regulatory frameworks must be carefully negotiated. To ensure that the benefits of nano-robotics are achieved safely, ethically, and fairly, scientists, medical specialists, ethicists, legislators, and patients will need to collaborate.

Furthermore, we recognize that the path to mainstream healthcare integration of nano-robotics will be marked by constant learning and adaptation. Not only will technology play a role, but healthcare institutions, medical education and training, and public involvement will all need to adjust to accommodate these developments.

To summarize, cautious optimism prevails. The use of nano-robotics in healthcare could lead to better patient care, healthier outcomes, and possibly even a longer life expectancy. To enhance healthcare in a way that is inclusive, safe, and ethical—one that acknowledges and enforces ethical boundaries—a combination of scientific acumen, technical innovation, and collective drive will be required. Only then will we be able to realize this potential.

As a result, this examination of nano-robotics in healthcare concludes with a call to action - a call for continual invention, extensive debate, and collaboration. The message is an urgent plea to all those involved in healthcare to work toward a future in which nano-robotics represent the pinnacle of medical technology and a beacon of hope for a more egalitarian and better society. The healthcare system of the future will be built on the foundation we lay today, and we know that the smallest machines, nano-robots, will make a significant difference in that system.

Chapter 4: Artificial Organ Transplants – Scientific and Ethical Challenges

Introduction to Artificial Organ Transplants

The narrative of artificial organ transplantation is a new chapter in the dynamic history of medical science; it is a story of human ingenuity and endurance. The story of our journey into the field of artificial organs is more than a chronicle of medical achievement; it is a story of optimism and a display of our tenacity in defying nature's restrictions. As we go on this adventure, we will encounter technology, ethics, and the fundamental question of how to sustain and enrich human existence.

The history of the idea of replacing failing organs with artificial ones is as fascinating as the technology itself. Organ transplantation has always been a medical frontier, and this story begins with its early days. The growth of this field has been marked by both success and failure, with the underlying goal of raising the standard of living for all people. From the first successful kidney transplant to the development of mechanical hearts, the history of organ transplantation highlights our never-ending pursuit of solutions to some of medicine's most unsolvable issues.

Traditional organ transplants have shown to be fantastic, but they also present a difficult landscape. There are numerous obstacles in this area, including a scarcity of organ donors, the danger of organ rejection, and the have to take immunosuppressants for the rest of one's life. The technology of artificial organs has brought forth a new dawn here. Because of the development of artificial organs, which can be totally synthetic or bioengineered, the constraints of organ donation will no longer exist in the future.

Right now, the field of artificial organ technology is a rich tapestry of experimentation and discovery. Organ replacements suitable with the human body are on the horizon, thanks to advances in stem cell research, biomaterials, and 3D printing. This type of technology does more than merely duplicate biological processes; it rethinks them and offers alternatives that may one day be more effective than organic organs.

The breadth of potential applications for artificial organs is both surprising and deeply significant. Artificial kidneys that eliminate the misery of dialysis, bioengineered lungs that evolve and mature with their recipients, and artificial hearts that eliminate the risk of cardiac rejection are all potential future technologies. With each new invention, there are stories of people whose lives have been improved and who have been given new hope.

There is a significant obstacle on the road to artificial organ transplantation. These innovations present substantial logistical and ethical challenges. Several variables must be carefully studied, including accessibility and equity difficulties, the psychological effects of living with artificial organs, and the legal frameworks that govern these technologies.

We hope that by introducing readers to this topic, we can help them gain a better understanding of both the science and the broader implications of artificial organs. Investigating the viability of artificial organ transplants offers a look into a future in which technology advances enable us to overcome current challenges and reach new heights. A novel that simultaneously celebrates medical science's achievements and prompts us to consider the social and ethical consequences of this exciting new era.

Advancements in Artificial Organ Technology

Artificial organ transplantation transports us to a world where cutting-edge medical research and technological innovation collide to create something really astonishing. In this segment of our tale, we showcase advancements in artificial organ technology, a discipline that is always pushing the limits of what is possible by imaginatively rethinking and improving on biological processes.

The voyage of artificial organ technology exemplifies the amazing progress in biomedical engineering and materials science. Fundamentally, this field is concerned with creating artificial organs that can supplement or completely replace their real-life counterparts. In this field, a wide range of advancements have been achieved, ranging from bioartificial organs made from living cells to totally synthetic organs made from biocompatible materials.

The use of 3D printing technology is one of the most innovative advances in this industry. 3D printing, also known as additive manufacturing, allows for the layer-by-layer production of complicated structures such as organ replicas. This technique has made it possible to create anatomically precise, patient-specific organ replicas. Scientists are researching the feasibility of 3D printing organs such as kidneys, livers, and hearts using a combination of biomaterials and live cells.

Bioengineering has also advanced significantly. The goal of scientists developing bioartificial organs is to use living cells to create organ tissues or whole organs in a controlled environment. This method's goal, in addition to being mechanically functional, is to create physiologically active organs that can interact with the body's own systems. The use of stem cells, specifically induced pluripotent stem cells, has resulted in considerable advancements in this sector. It allows for the cultivation of organs that are genetically compatible with the recipient, reducing the likelihood of rejection.

External replacements are no longer the only focus of advances in artificial organ technology. New implantable devices and systems are being developed to augment or restore the function of existing organs, either by cooperating with them or by operating within them. For example, scientists are developing implantable devices to assist people with cardiac issues, as well as artificial pancreases to control insulin levels in diabetics.

Each new advancement in the realm of artificial organs brings with it an intriguing story of hope and opportunity. These are stories about forging new ground in medical research, bringing hope to those facing organ failure, and seeing a future in which organ donation is abundant and may save lives despite a donor scarcity.

Nonetheless, there remain technological, biological, and ethical barriers to overcome in addition to these advancements. It is a monumental job to ensure that artificial organs are accessible, reliable, and safe to use. As we study the most recent developments in artificial organ technology, we are constantly reminded of the delicate balance that must be maintained between pushing the frontiers of what is practical and preserving the safety of those involved.. Join us on this incredible journey through artificial organ technology to see how it will change healthcare for years to come and how we will overcome today's challenges.

Applications in Transplant Medicine

Artificial organ development is a watershed moment in the ever-changing area of transplant medicine, bringing up intriguing new routes of treatment for countless patients. This section of our inquiry delves into the diverse applications of artificial organs, emphasizing their revolutionary potential to tackle some of transplant medicine's most serious issues.

Artificial organs have a complex and important role in transplant treatment. The chronic paucity of donor organs has long been a problem in the industry, and this approach may offer a solution. Every year, many patients are placed on transplant waiting lists, and unfortunately, many of them die before a donor is found. The emergence of artificial organs, which could eliminate the need for human donors, could be a potential silver lining in this bleak circumstance. Many lives could be saved if this breakthrough significantly reduces transplant waiting periods.

In addition to addressing the issue of organ scarcity, artificial organs have the potential to change transplantation in general. Consider the numerous successful artificial organ transplant case studies; each one highlights the remarkable progress made in this field. Patients who have received organ transplants have not only been able to overcome the immediate threat to their lives, but they have also experienced an improvement in their quality of life following the treatment. These accomplishments mean more than just excellent health to the patients and their families.

Furthermore, by lowering the risk of organ rejection, which is a major barrier with traditional transplants, artificial organs have the potential to radically revolutionize the area of transplant medicine. Artificial organs reduce the likelihood of transplant rejection due to their biocompatibility and, in some cases, ability to be personalized to each patient's unique biological profile. Better transplant outcomes and the elimination of the need for lifelong immunosuppressive medicines, each with their own risks and side effects, may be conceivable as a result of this advancement.

What happens to one patient is only one example of how synthetic organs effect society as a whole. Organ allocation regulations and post-transplant care are just two areas where their revolutionary potential may have an influence. The dynamics of organ transplantation would change dramatically in a world where artificial organs were widely available, from the management of transplant processes to patient eligibility standards.

However, there are additional considerations and challenges to address when introducing artificial organs into transplant treatment. The ethical considerations about resource allocation and the availability of such technologies are on par with the technological and surgical challenges. Examining how artificial organs might be used in transplant therapy leads us through a landscape where creative technology meets a huge human need.

This article depicts a future in which technology breakthroughs have rendered traditional organ transplants obsolete by looking into the possibility of artificial organs in transplant treatment. People waiting for organ transplants have reason to be enthusiastic about the future because it promises a longer, better, and more satisfying life than they currently have. Thinking about these applications calls to mind the transformative potential of artificial organ technology in healthcare and human lives.

Ethical Considerations in Organ Transplantation

As we navigate the intriguing terrain of artificial organ transplantation, it is vital to address the interconnected ethical considerations. In this round of research, we will look closely at the ethical and moral quandaries that arise in the realm of artificial organ transplantation, as well as the technical advancements. The issues we face here put into question the fundamental principles of medical ethics, as well as our own personal beliefs and social standards.

Many major ethical concerns have recently surfaced in the context of the creation of artificial organs. One of the most pressing issues is equity and access. Who will benefit from these potentially curative technologies? In a society where artificial organs have the ability to save lives, the ethical prioritization of resource allocation becomes critical. There are valid concerns about healthcare unfairness and inequality if these technologies are only available to the wealthy or those who live in specific places. Equal access to artificial organ transplants is both a moral need and a logistical challenge.

Another area of ethical concern is the impact of artificial organs on our sense of self and humanity. The inclusion of synthetic or bioengineered organs into the human body raises philosophical issues about the nature of humanity due to the blurring of natural and artificial boundaries. How do we feel like ourselves when we have artificial components in our bodies? How will our perspective on life and health change if human organs are replaced with artificial ones?

The research, development, and ultimate use of artificial organs raises new questions about autonomy and consent. Artificial organ transplantation is a medically complicated and emotionally charged option that people must carefully consider. Patients undergoing informed consent for these surgeries will enter uncharted area as they attempt to appreciate the potential implications, dangers, and daily challenges of coping with an artificial organ.

The greater social and cultural contexts are important in addressing the ethical issues underlying artificial organ transplantation. Consider how people perceive and use technology in society, as well as how cultural values and beliefs impact these perceptions. Everyone, from patients and ethicists to legislators and the general public, must be involved in the debate over the widespread use of artificial organs in healthcare.

As we study the moral terrain around artificial organ transplantation, we take stock of our obligations and reflect on our own behaviors. We are reminded that medical breakthroughs are defined not just by their scientific triumphs, but also by how well they align with our moral compass as we navigate these ethical quandaries. As we investigate these ethical issues, we are encouraged to contemplate the profound implications of artificial organ transplants for our values, society, and humanity as a whole. This research is critical to understanding the scope of artificial organ transplantation, a field where scientific marvels meet ethical quandaries.

Challenges in Implementation and Acceptance

There have been significant developments in the field of artificial organ transplants, but there have also been great challenges to overcome in terms of both practical application and public perception. In this portion of our tale, we look at the difficulties that people experience while bringing artificial organ technology from the lab to the patient's bedside, including physical and mental impediments.

There are numerous technical and medical obstacles to the clinical implementation of artificial organ technologies. It is critical to ensure the long-term dependability and safety of these organs. Because the human body is always changing, artificial organs must be able to work in unison with it for extended periods of time with no visible disturbances. Thorough research on the long-term impacts of synthetic materials on the human body, as well as large-scale clinical studies, are required for this. Furthermore, due to the complexity and level of expertise required by the surgical procedures involved, medical workers undertaking artificial organ transplantation must have special training.

Beyond technical problems, the widespread deployment of artificial organs creates severe logistical concerns. These include the manufacturing processes, supply chain organization, and the healthcare system required to support these cutting-edge medicines. A fundamental issue, particularly in resource-constrained settings, is ensuring that artificial organs are effective, accessible, and cost-efficient.

Artificial organ transplants are well regarded by the medical community as well as the general public. Social and psychological constraints have a significant impact on the acceptance of this technology. Many people are concerned about having a synthetic or bioengineered organ replace a natural one. Several variables can influence how the public perceives and accepts the concept of "synthetic life," including worries about the compatibility of the human body with artificial organs, the implications for individual identity, and ethical problems.

Cultural beliefs and values also influence how different civilizations see artificial organ transplantation. In some communities, the assimilation of synthetic parts into the human body may raise ethical and theological concerns. Addressing these concerns necessitates deliberate and informed debate among community people and medical experts alike.

Examining these barriers to adoption and implementation reveals that the path toward artificial organ transplants is complicated, requiring not just science but also societal norms, ethics, and human psychology. To overcome these difficulties, ethicists, politicians, patient advocacy groups, the general public, and medical researchers and practitioners must all collaborate. In light of these issues, we can better understand the complexities of developing artificial organ technology and the importance of creating a secure platform for individuals to ask questions and share their viewpoints in order to promote medical innovation.

Regulatory and Policy Frameworks

The threads of regulatory and policy frameworks are critical in assuring the safe, ethical, and effective adoption of this new technology in the intricate tapestry of artificial organ transplantation. This segment of our investigation focuses on the critical role of regulatory organizations and the need for comprehensive policy development in the field of artificial organ transplantation. We face the complexities and challenges of building a governance system that can keep up with rapid technical breakthroughs while protecting public health and ethical standards as we navigate this landscape.

The regulatory environment surrounding artificial organ transplants is as complicated as the technology itself. One of the most difficult tasks is developing guidelines and criteria to ensure the safety and efficacy of artificial organs. Because these organs are at the vanguard of biological innovation, they frequently defy established medical categorizations, prompting the development of new regulatory mechanisms. Regulatory bodies must strike a balance between the need to enable medical progress and the need to safeguard people from potential hazards. This entails rigorous testing, approval, and monitoring processes, which are critical for maintaining the integrity and trust in healthcare systems.

Furthermore, the policies governing artificial organ transplantation go beyond basic regulatory compliance. They include broader ethical, legal, and social concerns. Policymakers must handle a wide range of challenges, from defining ownership and accountability in the case of bioartificial organs to ensuring fair access to these life-saving technology. Because artificial organs have the potential to considerably extend human life, regulations must take into account the long-term repercussions, such as the influence on healthcare costs, insurance schemes, and social resources.

Navigating the legal and ethical environment of organ transplantation is fraught with difficulties. Artificial organs, particularly those based on bioengineered tissues or cells, create significant difficulties with donor rights, patient permission, and the definition of life and death. Developing policies that address these issues while maintaining cultural, ethical, and personal values is a complex task that demands careful consideration and stakeholder participation.

In this part, we underline the need of working together to build regulatory and policy frameworks. Not only should lawmakers and regulatory bodies be involved in this collaboration, but so should medical researchers, healthcare professionals, ethicists, and patient advocacy groups. The goal is to establish a governance framework that not only stimulates innovation and progress in artificial organ technology, but also assures that these innovations are carried out in an ethical, equitable, and public-health-friendly manner.

We are reminded of the delicate balance between innovation and regulation, between the promise of new medical solutions and the need to keep the greatest standards of safety and ethics, as we investigate the regulatory and policy implications of artificial organ transplants. This investigation is a call to action for proactive and forward-thinking policy formation that recognizes the transformative potential of artificial organ technology while adhering to medical ethics and patient care standards.

Impact on Healthcare Systems

Artificial organ transplantation is a fascinating and hard topic that will surely impact healthcare systems around the world as more is learnt about it. Part two of our inquiry focuses on how the use of artificial organs will change healthcare management, patient care, and the overall quality of life for both doctors and patients.

The development of artificial organ technology has the potential to change the healthcare system in a number of fundamental ways. The method of organ transplantation is anticipated to have the greatest impact. Artificial organs have the potential to significantly alleviate the present constraints of organ donation, such as donor organ scarcity and the problems associated with determining organ compatibility. By drastically reducing transplant waiting times, this reform could relieve huge load on healthcare systems and save countless lives.

Furthermore, the use of artificial organs may result in a paradigm shift in the management of organ failure. Healthcare facilities are now saddled with the resource-consuming responsibility of caring for patients whose organs are failing, which typically necessitates lengthy hospital stays, intensive care, and complex prescription regimens. Because of their lower rejection rates and longer lifespans, artificial organs may provide more efficient and long-term therapeutic choices. As a result of this development, patients benefit and healthcare systems may be able to better spend their resources.

Healthcare professionals will also need to prepare for significant changes brought about by the use of artificial organs. It necessitates a reconsideration of how doctors learn and what they accomplish in the field. New surgical procedures, postoperative care customized to artificial organs, and the management of any specific risks associated with these devices will demand training for surgeons, doctors, and other medical workers. A significant investment in medical education and training is required for artificial organ technology to be adopted safely and successfully.

The use of artificial organs in healthcare raises broader issues about healthcare policy and infrastructure. Because artificial organ transplantation necessitates specialized operating rooms, recuperation areas, and diagnostic equipment, hospitals and other healthcare facilities may need to adapt to fulfill these demands. The distribution and use of artificial organs raises tough concerns of equity, access, and financing that must be addressed by healthcare administrators and politicians.

Investigating the implications of artificial organs on healthcare systems reveals that embracing this technology represents a far-reaching revolution that affects all aspects of healthcare, not just the scientific community. To ensure that the benefits of artificial organ technology are realized effectively and equally, healthcare legislation, processes, and practices must be reevaluated, as well as collaboration across sectors.

To summarize, artificial organs have huge potential to improve healthcare. It depicts a future in which healthcare systems are streamlined for efficiency and efficacy, patient care is improved, and creative solutions to organ transplantation challenges are found. When we consider this possibility, we are filled with hope for a future in which medical knowledge is pushed to its maximum in order to improve people's lives all across the world.

Case Studies: Innovations and Controversies

When researching artificial organ transplantation, it is critical to look at real-life case studies that demonstrate the advancements and disputes around this surgery. These accounts depict the practical applications, problems, and ethical debates surrounding artificial organ technology. Each case study, which acts as a mini-version of the larger story, examines the merits, downsides, and ongoing debates around artificial organ transplantation.

The successful application of artificial organ technology in these case studies has resulted in new advances. For example, descriptions of groundbreaking surgeries that have used synthetic organs to replace damaged natural ones, or accounts of people whose lives have been significantly enhanced after receiving bioengineered organs. These cases emphasize both the practicalities and the potentially life-changing benefits of artificial organ transplantation.

Stories of success abound in this section, but so do those that underscore the field's fundamental disputes and difficulties. This can include examples where ethical or medical concerns developed as a result of artificial organ transplants failing to function properly. Examples include circumstances in which the long-term viability of bioengineered organs has been called into question, or in which the body's reaction to a synthetic organ has resulted in unexpected consequences. It is critical to evaluate these possibilities in order to completely comprehend the potential benefits and drawbacks of artificial organ technology.

These case studies also take into account the ethical debates that arise as a result of the medical usage of artificial organs. Many discussions have erupted about the distribution of these sophisticated organs, the pricing and availability of these medical cures, and the moral implications of artificially prolonging human life. Because these ethical concerns mirror larger society attitudes and ideals, there must be a comprehensive strategy for deploying artificial organ technology.

By evaluating these case studies, the field of artificial organ transplantation is regarded in a multifaceted perspective. The stories of debate and challenge serve as a reminder of the importance of rigorous research and ongoing refinement in the development and implementation of this technology, while the stories of inventiveness underscore its enormous potential to alter lives.

Our understanding of artificial organ transplantation is primarily reliant on case studies. They demonstrate how difficult it is to thrive in this inventive sector and how people are always attempting new things by linking real-life situations to theoretical components of this technology. As we progress in our understanding of this game-changing medical innovation, reading these experiences should cause us to reflect on the great promise and weighty responsibility of artificial organ technology.

Future Directions in Artificial Organ Research

As we continue to investigate the story of artificial organ transplantation, our focus switches to the future, wondering where this inventive science will go from here and what possibilities lay ahead. This section of our investigation is more than just a prediction of what will happen; it is an examination of the current state of artificial organ research as well as its potential future directions, difficulties, and opportunities. This enables us to picture a future in which innovation and a strong desire to improve human health propel medical science to new heights.

The field of artificial organs has a bright future ahead of it, thanks to our increased understanding of human biology and the lightning-fast speed of technical advancement. We are on the cusp of a new era in which the efficacy and acceptability of organ transplants could be considerably enhanced by the development of more advanced and biocompatible artificial organs. Recent advancements in tissue engineering, stem cell research, and biomaterials show enormous promise for the construction of artificial organs that can compete, if not outperform, their real-life counterparts.

One of the most potential future prospects is the creation of bioartificial organs that blend biological and synthetic parts. With this combination method, we may one day create artificial organs that fit like a glove in the body, eliminating the need for immunosuppression for the remainder of one's life and the risk of rejection entirely. If this research continues, it may one day be able to regenerate organs from the patient's own cells rather than simply replacing them.

The utilization of a patient's own cells in a layer-by-layer method during organ manufacturing opens the door to solutions that are specifically tailored to each person's unique genetic makeup and physiological make-up.

There are still significant roadblocks to development in artificial organ research. There are numerous challenges to overcome, including restricted technology, ethical considerations, and the requirement to conduct large-scale clinical trials to ensure safety and effectiveness. For these cutting-edge technologies to be widely adopted in healthcare, changes in medical education and practice, as well as strong regulatory frameworks and revised healthcare policies, will be required.

Also, keep in mind that there will be social and ethical implications as artificial organ research progresses. Healthcare equity, cost, and accessibility to these cutting-edge medicines must all be carefully considered. If the field is to progress, changes must be more equal, moral, and in keeping with the larger goals of public health.

Thinking about where this field of artificial organs could go from here is a strong reminder of the human intellect's infinite potential and unquenchable thirst for information. This exploration into the future is a rallying cry for continued creativity, cross-disciplinary collaboration, and ethical vigilance. In doing so, we acknowledge that the goal of artificial organ research should be to create a future in which everyone has access to potentially life-saving treatments, rather than simply overcoming present technological limitations. This journey exemplifies medical research's tremendous ability to do more than merely cure; it has the power to change lives and usher in a new era of optimism and promise in healthcare's dynamic story.

Side Story Subsection: Artificial Organs in Popular Culture

This interesting offshoot of our main story explores the media's and literature's intriguing depictions of artificial organs in the context of organ transplantation. The idea of artificial organs has captivated storytellers, filmmakers, and audiences alike, despite its factual veracity, as this journey through popular culture demonstrates. Our hopes, concerns, and dreams for the future of medicine and mankind are mirrored in this narrative, which also reflects our technological ambitions.

Science fiction has a long history of including artificial organs, which are frequently used to represent sophisticated technology and the merging of human and machine identities. The wide spectrum of depictions of these organs in fiction shows how modern technology affects human ethics and identity, from miraculous lifesavers to sources of existential anguish. As an example, the concept of artificially enhanced individuals with superhuman skills is a common thread in science fiction stories, prompting readers to consider the ethics of such augmentations and the very definition of humanity.

The idea of artificial organs has expanded beyond fiction and into the public sphere, where it has been discussed extensively. People are captivated and engaged by the newest advancements in artificial organ research, which is frequently covered in documentaries, news articles, and popular science conversations. The public's views and expectations on the trajectory of medical research are significantly influenced by these depictions.

Nevertheless, there are ambiguities to the way artificial organs are portrayed in popular culture. Although these stories have the power to enlighten and motivate, they also have the potential to distort the reality of technology, which in turn can cause misunderstandings and unreasonable expectations. The serious scientific and ethical concerns surrounding the creation and use of artificial organs may be overshadowed by the dramatic treatment of these technologies in certain works of fiction.

We incorporate components like media-based quizzes on artificial organs, conversations about how well these depictions mirror actual scientific knowledge, and made-up scenarios influenced by these stories to make this examination of these narratives more participatory and interesting. Readers are encouraged to critically engage with the portrayal of artificial organs in popular culture through these interactive components, which entertain as well.

As we delve further into the topic of artificial organ transplantation, this fascinating side narrative of these devices in popular culture becomes even more apparent. It helps connect people with scientists while also drawing attention to how cultural narratives shape our perceptions of and comfort with cutting-edge medical technology. We are reminded of the influence of storytelling as we make our way through these fictional and factual depictions of science and technology, which in turn reflect our hopes and fears for the future.

Conclusion: Navigating the Future of Artificial Organ Transplants

Our investigation into artificial organ transplantation is coming to a crossroads where looking back and forward are complementary ideas. In this final segment, we will take a reflective look at the path that lies ahead, rather than just restating the previous discussions. In doing so, we recognize both the progress that has been achieved and the difficulties and obligations that still lie ahead in the realm of artificial organ transplantation. With ethical thought, ongoing innovation, and collaborative effort as our guiding principles, we stand on the brink of a future that is ripe with opportunity.

There is no denying the revolutionary potential of artificial organ transplantation in the healthcare industry. There are innumerable people who are waiting for transplants that could save their lives, and this technique gives them hope that the scarcity of organ donors will one day no longer be a fatality. More individualized, efficient, and widely available treatments are becoming possible because of developments in artificial organs, which range from bioengineered tissues to completely synthetic substitutes.

However, we are well aware of the intricacies that come with such revolutionary advances as we look into the future. Equally important as the technical hurdles are the ethical concerns that arise from artificial organ transplantation, such as issues of equity and accessibility and the consequences for human identity. Everyone from doctors and ethicists to lawmakers and the general public will need to work together to find a way through these murky ethical waters.

Furthermore, healthcare systems across the globe will have to undergo revisions and changes due to the introduction of artificial organ technology. Not only that, but it also encompasses alterations to healthcare policy, infrastructure, and professional training, as well as the incorporation of new medical technology. Going forward, we need to take a comprehensive approach that thinks about the social and systemic effects of artificial organ transplantation in addition to the medical and scientific ones.

We conclude by considering how critical it is to keep pushing the boundaries of artificial organ technology through research and development. More research, stricter testing, and an emphasis on better patient outcomes are necessary to bring artificial organs closer to their full potential. In order to overcome the upcoming technical, ethical, and logistical obstacles, it will be crucial to work together across fields and countries.

A cautious hope permeates this investigation into artificial organ transplantation. Progress in this area is evidence of how far humanity has come in its quest for better health care. Nevertheless, achieving the potential of artificial organs necessitates not just technical expertise, but also a strong dedication to ethical values, fair healthcare, and the enhancement of human existence.

We believe that all those involved in healthcare will help usher in a future where artificial organ transplantation is more than a technological marvel; it will be a symbol of hope, equality, and compassion as we say goodbye to this chapter.

Introduction to the Microbiome

The microbiome is a large, nearly unseen area within the human body that is alive and vital to our survival. In this introduction, we will look at the microbiome, which is a secret universe of bacteria that lives within each of us and has a huge impact on our health, pleasure, and behavior. As we begin this exploration, we enter a strange and necessary world by revealing the delicate interaction between these microscopic residents and their human hosts.

Although the history of the microbiome is recent in medical history, it has swiftly become an important aspect of how we understand human health. The "microbiome" is the genome of all microorganisms (bacteria, fungi, viruses, and so on) that live in or on humans. Microbial communities can be found everywhere, from the interior of your digestive tract to the surface of your skin.

For much of medical history, these bacteria were largely ignored due to their reputation as disease-causing pathogens. However, with the development of advanced genetic sequencing technologies, our perspective on these bacteria has shifted dramatically. Even though some microbes can cause illness, modern science demonstrates that the vast majority of them are safe or even beneficial, and they play critical roles in digestion, immune system regulation, and neurological system function.

The human microbiome is astounding in its diversity and complexity. Individuals' microbiomes are shaped by factors such as genes, nutrition, environment, and lifestyle. This particular environment, like an organ, has a variety of effects on our health and physiology. The precise balance of microorganisms can be the difference between good health and sickness.

Changes in the composition of our microbiome have been linked to a variety of health issues, according to new research. These include both physical and mental health issues, such as obesity, diabetes, and inflammatory bowel disease, as well as mental health disorders such as anxiety and depression. These findings have cleared the door for additional research into the microbiome's potential significance in the treatment and prevention of various diseases.

When we investigate the microbiome, we are confronted with a significant lot of mystery. Interactions between bacteria and their consequences on human health are a mystery that has yet to be solved. Investigating the function of each microbe, their interaction, and the overall impact on the human body is a cutting-edge scientific field.

This introduction to the microbiome lays the framework for an exciting and critical investigation of human health. On this journey, we will challenge long-held views about microbes, reconsider our relationship with these microscopic entities, and discover a plethora of new potential for future medical discoveries. The microbiome is a complex and vast ecosystem, and our quest to comprehend it may lead to fresh perspectives on health and illness.

The Human Microbiome and Its Composition

In this section, we will continue our investigation of the microbiome by learning more about the complicated makeup of the bacteria that live on and in our bodies. This inquiry, which is more than just a trip into its diversity, provides insight into the complicated relationship between the microbiome and human health and welfare. The human microbiome contains an astonishing diversity of bacteria, all of which contribute considerably to our bodily functioning.

The human microbiome is astonishingly varied. It contains a wide range of microorganisms, including bacteria, viruses, fungi, and protozoa, all of which live in distinct parts of the body and play vital roles in its functioning. The microbiota of our digestive tract has received the most attention. This microbial population aids in digestion, absorption of nutrients, and vitamin synthesis. It is crucial for creating an immune system that can ward off hazardous illnesses, in addition to impacting our metabolism and weight.

All of our bodies' microbial communities, including those in our skin, mouth, lungs, and genitalia, are critical. The skin microbiome, by its barrier-forming function, simultaneously prevents harmful germs from entering and promotes wound healing. Oral health and systemic illnesses such as diabetes and heart disease are linked by the oral microbiome. Just as the microbiomes of the urogenital and respiratory tracts perform distinct but complementary roles, they have distinct effects on reproductive and respiratory health.

A person's microbiome is similar to a fingerprint: it is unique to each individual due to a combination of factors such as heredity, age, nutrition, environment, and manner of life. Because of this individual variety, understanding the role of the microbiome in health and illness is much more difficult. It suggests that our microbiome and health interact in a very individualized way, which gives hope for future specialized medical techniques.

This intricate microbial community, however, is precariously balanced. Disruptions to the microbiome can cause dysbiosis, an imbalance that has been linked to a variety of health issues such as gastrointestinal diseases, allergic reactions, and mental health issues. Antibiotics, dietary changes, and lifestyle considerations can all induce these abnormalities.

Microbiome and Its Impact on Health

More investigation into the enigmatic human microbiome reveals its far-reaching effects on many aspects of our health. This study goes beyond our superficial understanding of the microbiome as a collection of bacteria to reveal its critical role in determining our overall health. The status of these microbial communities influences every element of our existence, creating a complicated feedback loop between human health and the microbiome.

The microbiome is important in two areas: digestive health and nutritional absorption. The gut microbiome, a complex population of bacteria and other microorganisms that dwell in our intestines, is essential for digestion, mineral absorption, and even vitamin production. When this balance in the digestive tract is disrupted, diseases such as inflammatory bowel disease, obesity, and irritable bowel syndrome can emerge. Microbiome diversity may be linked to an increased risk of metabolic illnesses, since accumulating data suggests that the composition of our gut microbiome has a significant impact on our metabolic processes and weight.

In addition to digestion, the microbiome has a significant impact on immune function. It aids in the distinguishing of benign substances and potential infections due to its crucial role in immune response modulation and training. An unbalanced microbiome can cause an overactive or underactive immune system, which can lead to a variety of problems, including greater susceptibility to infections, allergies, and autoimmune diseases.

The gut-brain axis describes the relationship between microbiota and mental health. New research suggests that the gut bacteria regulates brain chemistry and behavior, which has far-reaching consequences for cognitive capacity and emotional stability. According to new research, anxiety, despair, and even neurodegenerative diseases may have their roots in a flora imbalance in the gut.

However, the microbiota and health have a complex two-way relationship; it is not a one-way street. Changes to our microbiome caused by factors such as food, stress, medicine, and lifestyle choices can have an impact on our health. A lack of fiber and a diet high in processed foods, for example, can contribute to a decline in microbiome variety, but the use of probiotics and prebiotics can help restore it to a healthy level.

The study of the human microbiome extends beyond simply naming the many bacteria that exist in our bodies and the roles they play. We are also learning about the significant effects that these microscopic organisms have on human health. The microbiome is no longer regarded as a bystander, but rather as an active participant in the biological processes that comprise human bodies. Understanding the human microbiome is like embarking on a fantasy adventure into a hidden part of ourselves, one that holds the answers to health and disease mysteries that we have just scratched the surface of.

Diet and Lifestyle Influences on the Microbiome

Our most recent work into the human microbiome focuses on the major influence of food and lifestyle on these complicated microbial populations. We investigate the complicated relationship between our microbiome and our daily routines, including how our diet and lifestyle choices affect this vital ecosystem within us. Our biology is merely one piece of the puzzle that is the role of our behaviors and surroundings in developing the microbiome.

Our food has a significant impact on the composition and health of our microbiome. The foods we eat can support or disrupt a diverse and balanced microbial community, known as dysbiosis. Eat a variety of nutritious foods, including fruits, vegetables, and whole grains, to keep your gut flora in good form. These foods are high in prebiotics, which are fibers used by good bacteria for nourishment and growth. A microbiome that is malnourished and prone to pathogenic bacteria proliferation, on the other hand, is connected with a diet high in processed foods, sugar, and saturated fat.

Other components of lifestyle, aside from food, can have an impact on the microbiome. A person's level of physical activity, amount of sleep, stress levels, and even exposure to nature can all have an impact. Regular physical exercise, for example, has been linked to the health advantages of increased microbial diversity. Similarly, obtaining enough sleep and regulating stress are critical for maintaining a healthy microbiome, as difficulties with circadian rhythms and chronic stress can lead to microbial population imbalances.

Furthermore, our modern way of life, which involves living indoors, being too clean, and constantly using antibiotics, has a significant impact on the microbiome. Although antibiotics save lives, they also kill dangerous microbial groups and permanently alter the microbiome. Similarly, a loss in microbiome diversity caused by a lack of exposure to diverse natural habitats, which is frequent in urban settings, may have an impact on our immune system and disease susceptibility.

The intricate network of interactions between bacteria, food, and lifestyle is always evolving. The bacteria in our bodies are strongly affected by the decisions we make in this never-ending cycle, which in turn affects our well-being in a variety of ways. It is critical to understand this link in order to improve health through specific dietary and lifestyle adjustments.

The more we understand about how our diet and habits influence the microbiome, the more control we have over this internal ecology. The health of the billions of bacteria that reside within us is determined by daily decisions like what we eat, how active we are, and how we deal with stress. This segment of the story implies that we should stop thinking of ourselves as isolated individuals and instead think of ourselves as part of a wider ecosystem, where caring for our microbiome is vital to caring for our general health.

The Future of Microbiome Research

Our focus shifts to the horizon as we dive deeper into the fascinating microbiome, wondering what the future of microbiome research may hold. We're not simply speculating in this section of our journey; we're looking ahead to the microbiome and the transformative changes it could bring to healthcare and public health. As our understanding of the microbiome grows, it has the potential to totally transform how we think about health and illness.

There is a lot of hope for the future of microbiome research, which could lead to a greater understanding of the complicated interplay between these bacterial communities and human health. One of the most potential future directions is the development of microbiome-based tailored medicine. With further research into the microbiome's role in health and illness, tailored treatment based on microbiome profiles is becoming a genuine option. This may lead to more tailored and efficient treatments for metabolic diseases and mental health disorders.

New methodologies and technologies for studying the microbiome are being developed. Researchers now have more information about the intricacy of the microbiome than ever before, thanks to advancements in genetic sequencing, bioinformatics, and artificial intelligence. These techniques offer light on how different microbial species interact with one another and how this impacts human physiology by processing huge amounts of data.

The study of microbiota has numerous applications. Investigating the gut-brain axis and what it means for mental and neurological illnesses is an exciting area of research. Research investigating the effects of gut microbiota on brain function may yield novel treatments for mental health conditions such as autism, anxiety, and depression. Similarly, dermatology stands to profit from research of the skin microbiome, which could lead to better therapies for eczema and acne.

Another intriguing emerging field of microbiome research is the development of microbiome-based therapeutics. This technique includes fecal microbiota transplantation (FMT), probiotics, and prebiotics. These drugs have the potential to cure a wide range of disorders, from autoimmune diseases to gastrointestinal problems.

The route forward in microbiome research, however, is not without challenges. When dealing with ethical issues, especially those concerning data privacy and the use of individual health records, caution will be required. Not only must new scientific information be created, but new initiatives in education, logistics, and legislation are also required to translate research findings into clinical practice.

As we consider where microbiome research will go from here, a mixture of hope and duty floods over us. As a result of advancements in this field, a more complete and tailored approach to healthcare may become the standard. The fact that we are diving into the future of microbiome research demonstrates how potential this topic is for improving our understanding of health and illness. A brighter future lies, one that holds the promise for a more thorough understanding of the interwoven web of life within each of us, as well as fresh treatment approaches.

Ethical and Privacy Considerations

We must address the ethical and privacy concerns that come with microbiome research as we move through its complex and ever-growing landscape. As we continue to learn more about the human microbiome, we face moral and ethical questions, and this section of our investigation isn't just filler for the scientific story. Research into the microbiome has many potential benefits for human health, but it also begs the issues of consent, privacy, and the moral use of individuals' health records.

Concerns about the security and privacy of individuals' medical records are at the forefront of microbiome research's ethical framework. Personal microbiome information can shed light on a person's health, habits, and maybe even genetic susceptibilities. Therefore, protecting personal data from misuse or illegal access is of the utmost importance. Data handling best practices in biomedical research are called into question by the need to implement strict rules and protections to ensure privacy when collecting, storing, and analyzing microbiome data.

Informed consent is another ethical consideration. Everyone who takes part in microbiome studies has a right to know what happens to their data, whether it's utilized for good or evil, and what their rights are in terms of privacy. When the field is still developing and the effects of the research on the future are unclear, this becomes even more crucial. The participants' autonomy must be respected throughout the process of getting consent, which must be comprehensive, open, and honest.

Additional ethical concerns arise from the potential effects of microbiome studies on healthcare policy and practice. There is increasing potential for the microbiome to impact healthcare decisions and public health efforts as our knowledge of its function in health and illness expands. Because of this, we need answers about how to fairly incorporate microbiome research into healthcare so that people of all income levels can reap the benefits of this field's advancements.

Personalized medicine, predictive analytics, and microbiome research all pose new ethical challenges, such as the possibility of genetic discrimination in health insurance or the workplace. The advancement of scientific knowledge and the protection of individual rights and liberties must be carefully balanced in order to navigate these difficulties.

Examining the privacy and ethical concerns surrounding microbiome research brings to light the fact that the path to scientific understanding involves more than simply advancing technology; it also involves negotiating the intricate terrain of human values and ethics. This investigation is an impassioned plea for a responsible and morally sound strategy to study the microbiome, one that protects people's right to privacy and autonomy while working to realize the field's immense promise for health benefits. The importance of strong governance, constant discourse, and ethical awareness in the ever-changing field of microbiome research is highlighted as we explore these ethical concerns, recognizing the weight of scientific knowledge and the responsibility that accompanies it.

Microbiome Therapies and Innovations

Here, we move from theoretical considerations of the microbiome to the creation of novel treatments and insights into its complex dynamics, the domain of actual applications. In this part of our journey, we delve into the groundbreaking cures that have come from microbiome research. We'll see how these advances are changing the way we approach healthcare. In this story of metamorphosis, we learn how to improve human health by utilizing the information found in our microbe friends.

Microbiome therapies are at the cutting edge of medical innovation, providing novel approaches to treating a wide range of illnesses. To name one of the most famous uses, probiotics are live bacteria that, when eaten, have positive effects on health. When the microbiome is out of whack, as can happen after antibiotic treatment or with gastrointestinal problems, probiotics can help get things back in balance. They improve the gut microbiome's general function while simultaneously restocking good bacteria and blocking dangerous microbes.

A further innovative field is the utilization of prebiotics, which are compounds that nourish good bacteria in the digestive tract. Foods high in fiber, which commonly contain prebiotics, aid in maintaining a balanced microbiome, which in turn improves gut health and may have an effect on general wellness. More and more people are looking for ways to improve their microbiome health by combining prebiotics and probiotics, a practice known as synbiotics.

Fecal microbiota transplantation is a more direct and occasionally contentious method of microbiome therapy (FMT). The goal is to restore a healthy microbial community in the gut by transferring stool from a healthy donor to a patient. In addition to inflammatory bowel disease and obesity, FMT has demonstrated efficacy in treating infections such as Clostridioides difficile.

Even beyond these treatments, there is a wealth of untapped potential in microbiome research's innovation pipeline. Personalized probiotic treatments derived from individual microbiome profiles, microbiome-based diagnostics that can identify diseases based on microbial signatures, and engineered bacteria that can transport therapeutic agents directly to the site of disease are all being investigated by researchers.

Nevertheless, there are many obstacles to overcome on the way to incorporating these medicines into conventional medicine. Thorough evaluations of safety and ethical considerations are necessary for scientific validation. To guarantee the safe and effective use of these innovative medicines, clear standards are required, as the regulatory environment is still changing.

We are seeing a tidal change in medicine as we delve into microbiome therapies and discoveries, moving our attention from diseases to the body's ecosystem. Recognizing the complex relationships between our microbiome and our health, these therapies offer a more comprehensive perspective on wellness. Investigating these treatments brings to light the limitless possibilities of our own biology, the efficacy of manipulating the microbiome for therapeutic purposes, and the ever-present necessity for creativity, precision, and morality in the quest for uncharted territories in medicine.

Case Studies: Success Stories and Challenges

As the story of microbiome research and its applications develops, it is crucial to look at real-life case studies that show the ups and downs of this area. In this part of our investigation, we will look at detailed examples of both the successes and failures of microbiome research in terms of its impact on human health. Every case study provides a concrete example of the possibilities and constraints of microbiome research, shedding light on its real-world applications in healthcare.

The incredible success stories in microbiome therapy are included in these case studies. Examples include accounts of people whose fecal microbiota transplantation (FMT) have been a game-changer in the treatment of recurring Clostridioides difficile infections. These achievements demonstrate the effectiveness of microbiome therapy and emphasize the profound difference they can make for patients who have tried everything else without success.

However, the difficulties and debates surrounding microbiome research are also highlighted by these case examples. For a thorough comprehension of the topic, it is essential to examine cases where microbiome therapies did not produce the desired results or where they caused unexpected consequences. Consideration should be given to issues such as the need to standardize FMT protocols, the safety of donor feces, and the long-term impact of changing the microbiome. All the more reason to conduct thorough scientific validation of medicines based on the microbiome, as these obstacles highlight how complicated the microbiome is.

The case studies also explore the regulatory and ethical hurdles that microbiome research has to overcome. Developing regulatory frameworks for new therapy types, resolving privacy concerns in microbiome data gathering, and managing the ethical implications of employing human-derived materials in therapeutics are all part of this. These problems show how important it is to strike a balance between encouraging innovation and protecting patients' rights and maintaining ethical standards.

We obtain a multi-faceted view of microbiome research by analyzing these case studies. The sector has the ability to revolutionize healthcare by providing innovative answers to long-standing medical problems, and the success stories prove it. While these obstacles highlight the need for continuous research, cooperation, and discourse, they also serve as a timely reminder of the significance of making ethical and careful progress in this area.

Our knowledge of microbial science relies heavily on these case studies. They illustrate the practical effects, both beneficial and detrimental, of microbiome research on healthcare and medicine, putting the theoretical and experimental parts of the topic into perspective. As we explore these experiences, we are prompted to reflect on the future of microbiome research, recognizing both its great promise and the obligations that come with it.

The Microbiome in Popular Culture and Media

In this fascinating offshoot of our microbiome investigation, we investigate the media's and the public's portrayal of this intricate web of microbes. The increasing interest in the microbiome and its impact on popular knowledge and imagination is shown here via its appearances in film, literature, and public discourse. Our fascination with and understanding of the invisible microbial environment within us and its effects on our well-being and daily experiences are mirrored in the ways the microbiome is portrayed in popular culture.

Despite its origins in the scientific community, the microbiome has recently gained widespread interest and is making appearances in popular culture. The microbiome is frequently portrayed in media as an enigmatic and potent internal force that can impact our mental and behavioral states in addition to our physical health. Despite the occasional exaggeration, these depictions have been essential in bringing greater attention to the microbiome and its impact on our health.

Fiction and cinema have drawn on the microbiome to tell stories that probe the limits of human and microbial interaction. Investigating the effects of microbial imbalances or manipulating the microbiome for better abilities in fiction frequently leads to deep philosophical and ethical inquiries. They hope that viewers would think about what can happen if humans interfere with these intricate ecosystems.

Articles and blogs that provide guidance on how to cultivate one's microbiome for improved health have also contributed to the microbiome's rise to prominence in the wellness and lifestyle sector. Although this has helped increase awareness of the microbiome, it has also spread false beliefs and simplistic solutions. Striking a balance between popular belief and fact in health trends pertaining to the microbiome is difficult.

More generally, how the media and popular culture depict the microbiome reflects how people feel about health, science, and the environment. It highlights how our bodies are increasingly understood to be interdependent on the microbial world and how this knowledge may shed light on health and illness.

The influence of narrative and public discourse on our understanding and beliefs is brought to light as we investigate the microbiome's visibility in the media and popular culture. This exploration highlights the need for responsible and accurate representation of scientific topics in popular media, ensuring that the fascination with the microbiome is grounded in a solid understanding of its true complexity and significance. When viewed through this perspective, the microbiome becomes clear that it is more than a mere scientific topic; it is also a social and cultural phenomena that has captured and holds the interest of the general people.In this fascinating offshoot of our microbiome investigation, we investigate the media's and the public's portrayal of this intricate web of microbes. The increasing interest in the microbiome and its impact on popular knowledge and imagination is shown here via its appearances in film, literature, and public discourse. Our fascination with and understanding of the invisible microbial environment within us and its effects on our well-being and daily experiences are mirrored in the ways the microbiome is portrayed in popular culture.

Despite its origins in the scientific community, the microbiome has recently gained widespread interest and is making appearances in popular culture. The microbiome is frequently portrayed in media as an enigmatic and potent internal force that can impact our mental and behavioral states in addition to our physical health. Despite the occasional exaggeration, these depictions have been essential in bringing greater attention to the microbiome and its impact on our health.

Fiction and cinema have drawn on the microbiome to tell stories that probe the limits of human and microbial interaction. Investigating the effects of microbial imbalances or manipulating the microbiome for better abilities in fiction frequently leads to deep philosophical and ethical inquiries. They hope that viewers would think about what can happen if humans interfere with these intricate ecosystems.

Articles and blogs that provide guidance on how to cultivate one's microbiome for improved health have also contributed to the microbiome's rise to prominence in the wellness and lifestyle sector. Although this has helped increase awareness of the microbiome, it has also spread false beliefs and simplistic solutions. Striking a balance between popular belief and fact in health trends pertaining to the microbiome is difficult.

More generally, how the media and popular culture depict the microbiome reflects how people feel about health, science, and the environment. It highlights how our bodies are increasingly understood to be interdependent on the microbial world and how this knowledge may shed light on health and illness.

The influence of narrative and public discourse on our understanding and beliefs is brought to light as we investigate the microbiome's visibility in the media and popular culture. This exploration highlights the need for responsible and accurate representation of scientific topics in popular media, ensuring that the fascination with the microbiome is grounded in a solid understanding of its true complexity and significance. When viewed through this perspective, the microbiome becomes clear that it is more than a mere scientific topic; it is also a social and cultural phenomena that has captured and holds the interest of the general people.

Conclusion: The Future of Health and the Microbiome

We have reached a turning point in our quest to comprehend the deep connection between these microbe populations and human health as we wrap up our exploration of the microbiome, a complex and intriguing world. This final section serves as both a retrospective on the investigated and an outlook on the future of microbiome research and its medical and health-related applications, including both the opportunities and threats that may arise. We are entering a new age in healthcare, where the microscopic organisms already present in our bodies will have a far greater impact on our health, happiness, and maybe even our survival.

More and more, it seems that our knowledge of the microbiome will determine the course of health in the future. Microbiome research has enormous potential to transform healthcare. We are opening doors to new treatments and prevention measures as we gain a better understanding of how these microbial communities impact our immune system and mental health. An intriguing avenue for customized medicine that could benefit from microbiome analysis is the possibility of customizing dietary recommendations and therapies to each person's own microbiota.

Nevertheless, there will be obstacles along the way. There is a great deal that remains unknown about the microbiome because of its intricate nature and the myriad ways it interacts with the human body. Ethics, rigor, and subtlety must characterize future research in this area. Ethical considerations, such as data privacy, permission, and the possible exploitation of personal health information, will become more pressing as our understanding of the microbiome grows.

In addition, a multidisciplinary strategy incorporating microbiologists, doctors, dietitians, mental health experts, and legislators will be necessary to bring microbiome research into clinical practice. The complex relationship between our internal microcosm, our external environment, and our way of life necessitates a holistic view of health.

Finally, we acknowledge that public discourse and education have a significant role in determining the trajectory of microbiome research and its practical applications. There must be truthful, balanced, and evidence-based reporting on the microbiome since it is gaining traction in the media and popular culture. As a result, people will be able to turn their increasing curiosity about the microbiome into well-informed decisions.

Future research on the relationship between the microbiota and health has great promise. This story takes a fresh look at old ideas about health and illness, encourages us to accept the nuanced nature of our biological connections, and paints a picture of a future where our tiniest friends play a key role in our general prosperity. We have just scratched the surface of what the microbiome has to offer in terms of potential for improving human health, and our exploration of this fascinating topic is far from done.

Closing Words

The third volume of "Discover Together: Storytelling for the Whole Family" has been an illuminating voyage into the complex and ever-changing world of health and biotechnology, and I want to thank you from the bottom of my heart for coming along for the ride. The journey into the human body, the wonders of modern medicine, and the limitless potential of biotechnology has been both enlightening and fascinating.

This volume's chapters cover a wide range of topics, from the latest innovations in artificial organ transplants to the fascinating microbiome. Its goal is to educate readers while simultaneously making them wonder at the complexity and power of the human body and the science that aims to improve our health. It is my sincere wish that you have gained knowledge and comprehension from this expedition, and that you have been intrigued and inspired by the intriguing relationship between biotechnology and health.

I hope that when you finish reading this book, it will make you wonder about the world, change the way you think about health and science, and make you want to learn more. Whether you're just starting to explore these subjects or are an expert, I hope that "Discover Together: Storytelling for the Whole Family" has been a pleasant and helpful guide.

You are my companion on this journey, and I am grateful. May there be an endless stream of new information, an abundance of shared experiences, and the pursuit of wisdom. Until we cross paths again, may your insatiable curiosity take you places no one has gone before.

Science Story Teller Dr. Byeong Jae "Ben" Chun has a background in chemical engineering, a love of storytelling, and a PhD. He has been a researcher for over 15 years. Ben knows firsthand the difficulties and rewards of inspiring a lifelong curiosity in science in his own children because he is a dad. Ben is able to fascinate audiences of all ages by simplifying difficult scientific ideas into stories with interesting plots and a lighthearted tone. Science Story Teller uses relatable stories, metaphors, and comedy to make science more approachable and exciting, with the hope of inspiring the future generation of scientists by piqueing their interest.

Experience the Exciting World of Health and Biotechnology in the Third Volume of "**Discover Together: Storytelling for the Whole Family.**" Discover the most recent developments in the field of medicine, including the microbiome's critical role in human health and the revolutionary possibilities of artificial organs.

A PhD in Chemical Engineering wrote this book, which is great for families since it simplifies complicated scientific findings into interesting stories. Brings to light the marvels and problems of modern biotechnology, this thought-provoking read is sure to inspire curiosity and discussion.

Anyone interested in the technological developments that are influencing our health and the future of medicine should read this book.

ISBN 9798878386418

The Art Of Mediating Gracefully

by Michael Lodge